INTERNATIONAL CENTRE FOR MECHANICAL SCIENCES

COURSES AND LECTURES - No. 280

# SINGULAR PERTURBATIONS IN SYSTEMS AND CONTROL

EDITED BY

M.D. ARDEMA

NASA
AMES RESEARCH CENTER

Springer-Verlag Wien GmbH

Originally published by Springer Verlag Wien-New York in 1983

ISBN 978-3-211-81751-3 ISBN 978-3-7091-2638-7 (eBook)
DOI 10.1007/978-3-7091-2638-7

## PREFACE

Over the past decade, singular perturbation theory has been applied to problems in systems and control. The singular perturbation approach has resulted in increased insight into dynamical and other systems and has led to high-efficiency computational algorithms. Consequently, the field has rapidly expanded to become one of the most active research areas in systems and control.

The 1982 Seminar on Singular Perturbations in Systems and Control brought together many of the leading researchers in singular perturbations for the purpose of providing a current, comprehensive review of the subject. The seminar was co-sponsored by the International Centre for Mechanical Sciences (CISM), the International Federation of Automatic Control (IFAC), the Italian National Research Council, the United Nations Educational Scientific and Cultural Organization (UNESCO), and the National Aeronautics and Space Administration (NASA). It was held at the CISM facility in Udine, Italy.

The Program Committee for the Seminar consisted of M. Ardema, P. Kokotovic, and R. O'Malley. The 24 attendees (representing ten different countries) heard thirteen invited lectures and participated in many fruitful discussions.

This volume is composed of the collected papers from the 1982 Udine Seminar. The types of systems considered in these papers include linear and nonlinear, discrete and continuous, and stochastic and deterministic. Both ordinary and partial differential equations are covered. The emphasis is on a concise statement of the existing theory. It is hoped that this document will provide a useful reference for both the theorist and the practicioner interested in singular perturbations in systems and control.

Finally, I would like to express my graditude to the CISM staff, and particularly to Paolo Serafini, for their hospitality and their diligence in organizing and conducting the Seminar.

Mark D. Ardema, Editor
October, 1982

## CONTENTS

## LIST OF CONTRIBUTORS

M.D. ARDEMA – M.S. 210-9, NASA-Ames Research Center, Moffett Field, CA 94035, U.S.A.

M. BALAS – Dept. of Electrical, Computer and Systems Engineering, Rensselaer Polytechnic Institute, Troy, NY 12181, U.S.A.

A. BENSOUSSAN – INRIA-Laboria, Domaine de Voluceau, Rocquencourt – B.P. 105, 78150 – le Chesnay, France.

C.M. BRAUNER – Laboratoire de Mathematiques, et d'Informatique, Ecole Centrale de Lyon, 36, route de Dardilly, B.P. No. 163, 69130 Ecully, France.

J.H. CHOW – Electric Utility Systems Engineering, General Electric, Schenectady, New York 12345, U.S.A.

P. HABETS – Institut de Mathematique Pure et Appliquee, Chemin du Cyclotron 2, B-1348 Louvain la Neuve, Belgium.

H. KHALIL – Dept. of Electrical Engineering and Systems Science, Michigan State University, East Lansing, MI 48824, U.S.A.

P.V. KOKOTOVIC – Coordinated Science Laboratory, University of Illinois, Urbana, Ill. 61801, U.S.A.

J.P. QUADRAT – INRIA-Laboria, Domaine de Voluceau, Rocquencourt – B.P. 105, 78150 – le Chesnay, France.

R.E. O'MALLEY – Dept. of Mathematics, Rensselear Polytechnic Institute, Troy, NY 12181, U.S.A.

A. VAN HARTEN – Mathematical Institute, State University of Utrecht, P.O. Box 80010, 3508 TA Utrecht, The Netherlands.

J.C. WILLEMS – Dept. of Mathematics, University of Groningen, P.O. Box 800, 9700 AV Groningen, The Netherlands.

B. NICOLAENKO – Center for Nonlinear Studies, M.S. 610, Los Alamos National Laboratory, Los Alamos, NM 87545, U.S.A.

# AN INTRODUCTION TO SINGULAR PERTURBATIONS IN NONLINEAR OPTIMAL CONTROL

M.D. Ardema
NASA Ames Research Center
M.S. 210-9
Moffett Field, Calif. 94035
U.S.A.

## 1. INTRODUCTION - SECOND ORDER UNCONTROLLED SYSTEM

### 1.1. Theory

Before turning to the general n-dimensional nonlinear optimal control problem, which is of primary interest in this paper, we will first consider a much simpler problem, namely the singularly perturbed, uncontrolled, autonomous, initial-value problem

$$\left.\begin{aligned} \frac{dx}{dt} &= f(x,y) \ ; \qquad x(\varepsilon,0) = x_o \\ \varepsilon\,\frac{dy}{dt} &= g(x,y) \ ; \qquad y(\varepsilon,0) = y_o \end{aligned}\right\} \tag{1.1}$$

where $x(\varepsilon,t)$ and $y(\varepsilon,t)$ are scalars, $\varepsilon < 0$, and $x_o$ and $y_o$ are constants.

This problem has been extensively studied in [1] and elsewhere; the present development is based on Chapter 4 of [2]. The case in which $f(\cdot)$ and $g(\cdot)$ are linear functions of x and y with constant coefficients has been analyzed by the method of matched asymptotic expansions in [3]. Analysis of the system (1.1) will provide a useful introduction to the treatment of nonlinear differential equations by matched asymptotic expansions.

We are interested in (1.1) under the assumptions that $\varepsilon$ is "small" (i.e. relative to the other parameters of the system). In this case, dy/dt may be large (compared with dx/dt) and we refer to x and y as the slow and fast variables, respectively.

When seeking a solution to (1.1), it is natural to set $\varepsilon = 0$ and solve the resulting problem with the hope that a reasonable approximation will be obtained. The problem with $\varepsilon = 0$ is called the reduced problem and its solution is denoted $x_r(\cdot)$ and $y_r(\cdot)$; i.e. these functions satisfy

$$\frac{dx_r}{dt} = f(x_r, y_r) \; ; \qquad x_r(0) = x_o$$
$$0 = g(x_r, y_r) \qquad (1.2)$$

Because this system is first order, only one initial condition can be met and it is natural to retain the condition on the slow variable and meet the condition on $y(\cdot)$ by allowing a discontinuity at $t = 0$. Of prime importance is the relation of the solution of the reduced problem to that of the full problem. The best that can be hoped for is that $x_r(\cdot)$ is a good approximation to $x(\cdot)$ everywhere and that $y_r(\cdot)$ is a good approximation to $y(\cdot)$ everywhere except near $t = 0$. The situation is shown in

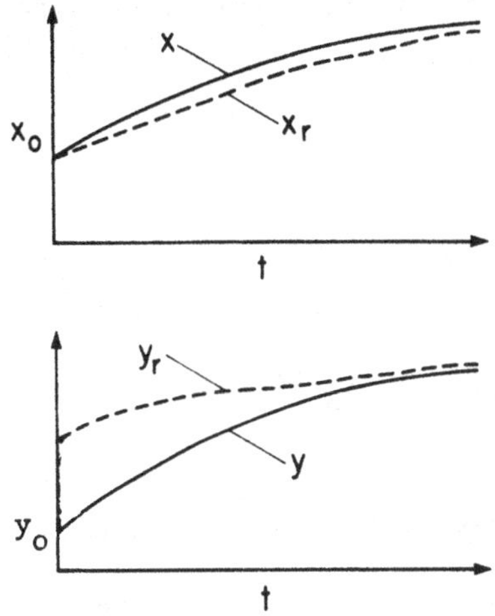


Figure 1.1. Sketch of full and reduced solutions.

in Figure 1.1. We assume that the exceptional case of $0 = g(x_o,y_o)$ is excluded.

To study the behavior of $y(\cdot)$ near $t = 0$, the time scale is "stretched" by introducing the transformation

$$\tau = \frac{t}{\varepsilon} \tag{1.3}$$

in (1.1) to get

$$\begin{aligned} \frac{dx}{d\tau} &= \varepsilon f(x,y) \; ; \qquad x(\varepsilon,0) = x_o \\ \frac{dy}{d\tau} &= g(x,y) \; ; \qquad y(\varepsilon,0) = y_o \end{aligned} \tag{1.4}$$

These are called the boundary layer equations. A reasonable approximation to these equations may, hopefully, be obtained by setting $\varepsilon = 0$, which gives

$$x_b = x_o$$
$$\frac{dy_b}{d\tau} = g(x_o, y_b) \; ; \qquad y_b(0) = y_o \tag{1.5}$$

Comparing (1.2) and (1.5) shows that the reduced solution evaluated at $t = 0$ is an equilibrium (singular) point of the zeroth order boundary layer equation (ZOBLE) given by (1.5). The crucial condition turns out to be the stability of the ZOBLE with respect to this equilibrium point. The following result is due to Tihonov [4]:

Theorem 1.1. Suppose that (A) $f(\cdot)$ and $g(\cdot)$ in (1.1) are continuous in some open region $\Omega$ of their domain, (B) both the reduced problem (1.2) and the full problem (1.1) have unique solutions on some interval $0 \le t \le T$, (C) there exists an isolated root $y_r = \phi(x_r)$ of (1.2) in $\Omega$, i.e., $0 = g(x_r, \phi(x_r))$, (D) the root $y_r = \phi(x_r)$ is an asymptotically stable equilibrium point of the ZOBLE (1.5), and (E) the initial point $(x_o, y_o)$ is in the domain of influence of this root. Then, $\lim_{\varepsilon \to 0^+} x(\varepsilon, t) = x_r(t)$ uniformly on $0 \le t \le T$ and $\lim_{\varepsilon \to 0^+} y(\varepsilon, t) = y_r(t)$ uniformly on any closed subinterval of $0 < t \le T$.

Further discussion and generalizations of Tihonov's theorem may be found, for example, in [1], and [2], [4]-[7]. An interesting geometrical interpretation of Tihonov's theorem is contained in [8] where the connection is made to the phenomenon of relaxation oscillations.

From a practical standpoint, the problem with Tihonov's theorem is that stability analysis of nonlinear systems is typically very difficult. First, there may be more than one root of $0 = g(x_r, \phi(x_r))$ and the relevant one has to be identified. Second, the local stability properties of the root must be determined. Finally, the domain of influence of the root must be established, usually the most difficult step.

Typically the best that can be done is a local analysis based on linearizing (1.5) about (1.2). This gives an expression for the behavior of the ZOBLE in the vicinity of its equilibrium point: namely,

$$\frac{dy_b}{d\tau} = \left.\frac{\partial g}{\partial y}\right|_{\substack{x=x_o \\ y=y_r}} (y_b - y_r)$$

Thus, it is necessary for the stability discussed in Theorem 1.1 that

$$\left.\frac{\partial g}{\partial y}\right|_{\substack{x=x_o \\ y=y_r}} < 0 \tag{1.6}$$

Since (1.6) is a local condition, it should be used with caution.

If a more accurate approximation to the solution of (1.1) than is given by the reduced solution is required, it is natural to attempt a solution by asymptotic expansion in the small parameter $\varepsilon$. Vasileva [5] has pioneered development of methods of solution of (1.1). In her method a solution is sought in the form of a sum of asymptotic expansions which approximates the solution both "inside" and "outside" the boundary layer. The central result of Vasileva's analysis is the following.

Theorem 1.2. If f(•) and g(•) are n+1 times continuously differentiable in some open region of their domain and if assumptions (B) through (E) of Theorem 1.1 hold, then there exist asymptotic series $\sum_i x_i(t)\varepsilon^i$, $\sum_i \bar{x}_i(\frac{t}{\varepsilon})\varepsilon^i$, $\sum_i y_i(t)\varepsilon^i$, $\sum_i \bar{y}(\frac{t}{\varepsilon})\varepsilon^i$, which are asymptotic expansions of the solution of the full system $x(\varepsilon,t)$ and $y(\varepsilon,t)$, that is

$$\left.\begin{aligned} &\left| x(\varepsilon,t) - \sum_{i=0}^{n} x_i(t)\varepsilon^i - \sum_{i=0}^{n} \bar{x}_i(\tfrac{t}{\varepsilon})\varepsilon^i \right| = O(\varepsilon^{n+1}) \\ &\left| y(\varepsilon,t) - \sum_{i=0}^{n} y_i(t)\varepsilon^i - \sum_{i=0}^{n} \bar{y}_i(\tfrac{t}{\varepsilon})\varepsilon^i \right| = O(\varepsilon^{n+1}) \end{aligned}\right. \tag{1.7}$$

in $\Omega$ for $0 < t \leq T$.

Vasileva's proof of this theorem in [5] is constructive in that an explicit method of obtaining these expansions is derived. We note that the smoothness required of the system functions depends on the order of the expansions being sought in the obvious way and that two-term expansions are generally needed for singular perturbation problems. In fact, in Vasileva's method the terms in the second expansions, such as $\bar{x}_i(\frac{t}{\varepsilon})$, actually consist of two terms themselves, the first of which is a solution of the boundary layer system and the second of which is a solution of a differential equation for the "common parts" (see next section for a definition of this term). We should also remark the Vasileva established her result for the more general case of vector x and vector y.

If a function $f(\varepsilon)$ has an asymptotic expansion $\sum_{i=0}^{\infty} a_i\varepsilon^i$ we will call

$\sum_{i=o}^{n} a_i\varepsilon^i$ the "n-th order approximation to $f(\varepsilon)$", in conformance with the mathematical literature on singular perturbation. In the fluid mechanics literature on asymptotic methods, $\sum_{i=o}^{n} a_i\varepsilon^i$ is sometimes called the "n+1-th order approximation to $f(\varepsilon)$". Because $|f(\varepsilon) - \sum_{i=o}^{n} a_i\varepsilon^i| = O(\varepsilon^{n+1})$, in our terminology the n-th order approximation is accurate to order n+1.

## 1.2. Asymptotic Analysis

We now procede to develope an asymptotic solution of the system (1.1). Instead of using Vasileva's method, briefly refered to in the previous section, we adapt the method of matched asymptotic expansions (MAE). This method was developed to solve certain nonlinear partial differential equations arising in fluid mechanics. Current expositions of the method are given in [9]-[12].

In the MAE method, seperate solutions are obtained for the region away from the boundary (outer region) and for the region near the boundary (inner region or boundary layer) by asymptotic expansion techniques. The unknown constants arising in the outer solution (the outer solution is not required to satisfy boundary conditions) are determining by matching the two solutions, i.e. requiring that they have the same behavior in an overlap region. The inner and outer solutions are then combined to give a uniformly valid asymptotic representation of the solution.

The MAE approach is adapted here because: (1) there has been a great deal of practical experience with the method (mostly in fluid mechanics, (2) it solves the inner and outer problems independently and there is often some practical use for these independent solutions, and (3) there is a great deal of flexibility available in constructing the uniformly valid solution. Because the method is theoretically equivalent to Vasileva's, we assume that the hypotheses of Theorem 1.2 hold.

Denote the outer solution of (1.1) by $x^o(\cdot), y^o(\cdot)$; since this solution describes the system behavior away from the boundary, $x^o(\cdot)$ and $y^o(\cdot)$ satisfy

$$\left.\begin{aligned} \frac{dx^o}{dt} &= f(x^o, y^o) \\ \varepsilon \frac{dy^o}{dt} &= g(x^o, y^o) \end{aligned}\right\} \qquad (1.8)$$

which is (1.1) without the boundary conditions. Upon denoting the inner solution by $x^i(\cdot), y^i(\cdot)$, (1.4) gives

$$\left.\begin{aligned} \frac{dx^i}{d\tau} &= \varepsilon f(x^i, y^i) \; ; \qquad x^i(\varepsilon, 0) = x_o \\ \frac{dy^i}{d\tau} &= g(x^i, y^i) \; ; \qquad y^i(\varepsilon, 0) = y_o \end{aligned}\right\} \qquad (1.9)$$

To solve (1.8), we express $x^o$ and $y^o$ in terms of asymptotic power series in $\varepsilon$:

$$\left.\begin{aligned} x^o(\varepsilon, t) &= x_o^o(t) + x_1^o(t)\varepsilon + x_2^o(t)\varepsilon^2 + \ldots \\ y^o(\varepsilon, t) &= y_o^o(t) + y_1^o(t)\varepsilon + y_2^o(t)\varepsilon^2 + \ldots \end{aligned}\right\} \qquad (1.10)$$

Putting these expressions in f(•) and g(•) and expanding about $\varepsilon = 0$ gives

$$\left.\begin{aligned} f(x^o,y^o) &= f(x_o^o+x_1^o\varepsilon+\ldots,y_o^o+y_1^o\varepsilon+\ldots) \\ &= f(x_o^o,y_o^o) + (f_{x_o^o}x_1^o + f_{y_o^o}y_1^o)\varepsilon + \ldots \\ &= f_o^o + f_1^o\varepsilon + \ldots \\ g(x^o,y^o) &= g(x_o^o,y_o^o) + (g_{x_o^o}x_1^o + g_{y_o^o}y_1^o)\varepsilon + \ldots \\ &= g_o^o + g_1^o\varepsilon + \ldots \end{aligned}\right\} \quad (1.11)$$

where, for example

$$f_{x_o^o} = \left.\frac{\partial f}{\partial x^o}\right|_{\substack{x^o=x_o^o \\ y^o=y_o^o}}$$

Putting (1.10) and (1.11) in (1.8) yields

$$\left.\begin{aligned} \frac{dx_o^o}{dt} + \frac{dx_1^o}{dt}\varepsilon + \ldots &= f_o^o + f_1^o\varepsilon + \ldots \\ \frac{dy_o^o}{dt}\varepsilon + \ldots &= g_o^o + g_1^o\varepsilon + \ldots \end{aligned}\right\} \quad (1.12)$$

Equating the coefficients of like powers of $\varepsilon$ gives the following sequence of problems for the coefficients of the series (1.10):

$$\frac{dx_o^o}{dt} = f_o^o; \qquad 0 = g_o^o \tag{1.13}$$

$$\frac{dx_j^o}{dt} = f_j^o; \qquad \frac{dy_{j-1}}{dt} = g_j^o \quad (j > 0) \tag{1.14}$$

Note that each of these problems is first order and that, except for the first one (1.13), they are linear. The constants of integration of these equations, say $C_j^o$, are as yet unknown.

To solve (1.9), put

$$\left.\begin{aligned} x^i(\varepsilon,\tau) &= x_o^i(\tau) + x_1^i(\tau)\varepsilon + x_2^i(\tau)\varepsilon^2 + \ldots \\ y^i(\varepsilon,\tau) &= y_o^i(\tau) + y_1^i(\tau)\varepsilon + y_2^i(\tau)\varepsilon^2 + \ldots \end{aligned}\right\} \tag{1.15}$$

Expanding about $\varepsilon = 0$ gives

$$\left.\begin{aligned} f(x^i,y^i) &= f(x_o^i,y_o^i) + (f_{x_o^i}x_1^i + f_{y_o^i}y_1^i)\varepsilon + \ldots \\ &= f_o^i + f_1^i\varepsilon + \ldots \\ g(x^i,y^i) &= g(x_o^i,y_o^i) + (g_{x_o^i}x^i + g_{y_o^i}y_1^i)\varepsilon + \ldots \\ &= g_o^i + g_1^i\varepsilon + \ldots \end{aligned}\right\} \tag{1.16}$$

Substituting (1.5) and (1.6) in (1.9) leads to

$$\left.\begin{aligned} &\frac{dx_o^i}{d\tau} + \frac{dx_1^i}{d\tau}\varepsilon + \ldots = f_o^i\varepsilon + \ldots \; ; \; x_o^i(0) + x_1^i(0)\varepsilon + \ldots = x_o \\ &\frac{dy_o^i}{d\tau} + \frac{dy_1^i}{d\tau}\varepsilon + \ldots = g_o^i + g_1^i\varepsilon + \ldots ; \; y_o^i(0) + y_1^i(0)\varepsilon + \ldots = y_o \end{aligned}\right\} \quad (1.17)$$

Equating the coefficients of like powers of ε gives the following sequence of problems:

$$\frac{dx_o^i}{d\tau} = 0; \; \frac{dy_o^i}{d\tau} = g_0^i; \; x_o^i(0) = x_o; \; y_o^i(0) = y_o \qquad (1.18)$$

$$\frac{dx_j^i}{d\tau} = f_{j-1}^i; \; \frac{dy_j^i}{d\tau} = g_j^i; \; x_j^i(0) = 0; \; y_j^i(0) = 0 \quad (j > 0) \qquad (1.19)$$

Each of these problems is of first order and, as before, all are linear except for the first one (1.18).

The constants of integration arising in the outer solution are determined by imposing a limit matching rule. This rule states that the asymptotic behavior of the outer solution when extended into the inner region must be the same as that of the inner solution when it is extended into the outer region. Formally, this means that the outer solution, when expressed in the inner variables and expanded for small ε, must agree with the inner solution, when expressed in the outer variables and expanded for small ε. Here we will write this limit matching rule as

$$\lim_{\substack{\varepsilon\to 0\\ t\to 0\\ \tau\to\infty}} \{x^o(\varepsilon,t) - x^i(\varepsilon,\tau)\} = 0 \qquad (1.20)$$

or (note that $\varepsilon \to 0$ faster than $t \to 0$):

$$\lim_{\substack{\varepsilon\to 0\\ t\to 0\\ \varepsilon/t\to 0}} \{x^o(\varepsilon,t) - x^i(\varepsilon,\tfrac{t}{\varepsilon})\} = 0 \tag{1.21}$$

The fast variable y will be matched if x is and thus it suffices to match only x.

For matching, we need the behaviour of $x^o(\varepsilon,t)$ for small t and $x^i(\varepsilon,\tau)$ for large $\tau$. To get the former, expand the coefficients of (1.10) in a power series about t = 0:

$$\left.\begin{aligned} x_j^o(t) &= x_j^o(0) + \left.\frac{dx_j^o}{dt}\right|_0 t + \ldots \\ \dot{x}_j^o(t) &= C_j^o + f_j^o(0)t + \ldots \end{aligned}\right\} \tag{1.22}$$

where (1.13) and (1.14) were used.

For $x^i(\varepsilon,\tau)$, note first that (1.18) reduces to

$$x_o^i(\tau) = x_o; \ \frac{dy_o^i}{d\tau} = g(x_o,y_o^i(\tau)); \ y_o^i(0) = y_o \tag{1.23}$$

By our assumption of asymptotic stability of (1.23), $y_o^i$ approaches the value $\phi(x_o)$ given by $0 = g(x_o,\phi(x_o))$, and hence

$$f_o^i(\tau) = f(x_o^i,y_o^i) = f(x_o,y_o^i)$$

approaches $f(x_o,\phi(x_o))$ as $\tau \to \infty$. To proceed further, knowledge of the functions $f(\cdot)$ and $g(\cdot)$ is needed. To illustrate the matching procedure we make the assumption that $f_o^i$ reaches its final value $f(x_o,\phi(x_o)) = f_o^i(\infty)$ in some finite time $\tau^*$ and remains there from then on, as illustrated in Figure 1.2. This would also correspond to the case of a numerical solution in which the integration is stopped when the solution becomes sufficiently close to its equilibrium value. Then from (1.19) the behavior of $x_1^i$ for large $\tau$ (i.e., $\tau \geq \tau^*$) is

$$x_1^i(\tau) = \int_0^\tau f(x_o,y_o^i(\eta))d\eta = I + f_o^i(\infty)(\tau - \tau^*)$$

$$x_1^i(\tau) = -I^* + f_o^i(\infty) \tag{1.24}$$

where

$$I^* = -\int_0^{\tau^*} f_o^i(x_o,y_o^i(\eta))d\eta + f_o^i(\infty)\tau^* \tag{1.25}$$

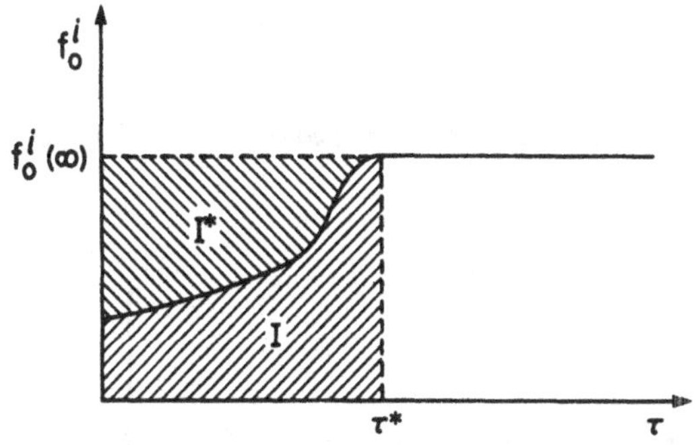


Figure 1.2. Sketch of $f_o^i(\tau)$ and definition of $I^*$ and I.

We are now ready to match. Put (1.22) in (1.10), put (1.23) and (1.24) in (1.15), and substitute the resulting expressions into (1.21) to get

$$\left.\begin{aligned} &\lim_{\substack{\varepsilon\to 0^+\\ t\to 0\\ \varepsilon/t\to 0}} \{C_o^o+f_o^o(0)t+C_1^0\varepsilon+\ldots-[x_o+(-I^*+f_o^i(\infty)\tfrac{t}{\varepsilon})\varepsilon+\ldots]\} = 0 \\ &\lim_{\substack{\varepsilon\to 0^+\\ t\to 0\\ \varepsilon/t\to 0}} \{C_o^o+f_o^o(0)t+C_1^o\varepsilon-x_o+I^*\varepsilon-f_o^i(\infty)t+\ldots\} = 0 \end{aligned}\right\} \quad (1.26)$$

Equating coefficients of like powers $t^k\varepsilon^\ell$ gives the sequence of conditions

$$\left.\begin{aligned} &C_o^o - x_o = 0 \\ &f_o^o(0) - f_o^i(\infty) = 0 \\ &C_1^o + I^* = 0 \\ &\vdots \end{aligned}\right\} \quad (1.27)$$

The first and third of these yield the first two unknown initial conditions as

$$\left.\begin{aligned} &C_o^o = x_o \\ &C_1^o = -I^* = \int_0^{\tau^*} f(x_o,y_o^i(\eta))d\eta - f_o^i(\infty)\tau^* \end{aligned}\right\} \quad (1.28)$$

The second of (1.27) is satsified trivially because of the assumed smoothness of $f(\cdot)$. The second of (1.28) is indentical to equation (40.23) of [1] and is a relation which was first derived in [5].

We have now obtained a representation of the solution near $t = 0$ (the inner solution) and a representation away from $t = 0$ (the outer solution). This leaves the somewhat awkward question of what is the appropriate value of t at which to switch from one to the other. To circumvent this problem and obtain a representation uniformly valid for all $t \geq 0$, a composite solution may be formed. There are many ways to form such a solution, but the most common is the additive composition

$$x^a(\varepsilon,t) = x^o(\varepsilon,t) + x^i(\varepsilon,\tfrac{t}{\varepsilon}) - CP(\varepsilon,t) \tag{1.29}$$

where CP is the "common part", i.e. the terms which cancel out in the matching. From (1.26) these are, to first order in $\varepsilon$,

$$CP(\varepsilon,t) = x_o + f_o^i(\infty)t - I^*\varepsilon. \tag{1.30}$$

The function $x^a(\cdot)$ has the properties that it satisfies the initial conditions, resembles the inner solution near $t = 0$, resembles the outer solution away from $t = 0$, and is smooth. For the problem under consideration,

$$\begin{aligned} x^a(\varepsilon,t) = x_o^o(t) + x_1^o(t)\varepsilon + \varepsilon \int_0^{t/\varepsilon} f(x_o,y_o^i(\eta))d\eta \; - \\ - \dot{f}_o^o(0)t + I^*\varepsilon + \dots \end{aligned} \tag{1.31}$$

The additive composite solution for y(t) is formed in the same way.

Our analysis of the relatively simple system (1.1) has revealed most of the features of the application of the method of matched asymptotic expansions to nonlinear ordinary differential equations. To sum up, following observations are made:

(1) Application of the method of MAE results in two sequences of problems, denoted the inner and outer, each of which is of lower dimension than the original system. These problems are linear, except for the first one in each sequence.

(2) To satisfy Theorem 1.2, asymptotic stability of the zeroth order inner problem is required. In the MAE method, this requirement arises naturally through matching relations such as (1.20). Higher order terms may be unbounded for fixed $\varepsilon$ as $\tau \to \infty$ as in fact, $x_1^i(\tau)$ is as seen in (1.24).

(3) The zeroth order outer problem is just the reduced problem obtained by setting $\varepsilon = 0$ and retaining the boundary condition on the slow variable x(•). For higher order outer problems, the appropriate initial conditions are not $x_j^o(0) = 0$, $j > 0$, as one might first suppose.

(4) The matching relations depend on the nature of the functions f(•) and g(•).

(5) To determine the unknown constants in the outer problems, only x(•) need to be matched; y(•) will then match identically.

(6) The matching relations give more conditions than there are free constants as can be soon from (1.26) and (1.27). In complex problems it may not be possible to satisfy all these relations. In such a case, more general expansions in $\varepsilon$ than power series and more general transformations than (1.3) have to be used which introduce more arbitrary parameters available to satisfy the matching conditions.

(7) The matching is not done on the basis of a one-to-one relation between like orders of the inner and outer solutions. For example, (1.26) shows that a term of the zeroth order outer solution has matched with a term of the first order inner solution.

(8) The common part, CP, needed for construction of a composite solution, is not simply equal to $x^o(\varepsilon,0)$ or to $x^i(\varepsilon,\infty)$ as may be seen from (1.30).

(9) The integral $I^*$, defined in (1.25), provides a useful quantitative measure of the magnitude of the boundary layer correction. As can be seen from Figure 1.2, it measures both quantities which determine this correction, namely the distance between the value of the outer solution at the boundary and the boundary value ($|y_r(0)-y_o|$) and the rate of decay of the boundary layer motion (determined by the value of $\varepsilon$).

(10) Because of our assumption of boundary layer asymptotic stability,

$$I^\infty = \lim_{\tau\to\infty} [- \int_0^\tau f_o^i(x_o, y_o^i(\eta))\eta + f_o^i(\tau)\tau] < \infty$$

and $I^* \to I^\infty$ as $\tau \to \infty$. However, the definition in (1.25) was adapted because it is more useful in practice.

### 1.3. Example

To illustrate the ideas of the previous section, the example given on page 275 of [1] will be solved to first order by the method of MAE. The example is solved in [1] using Vasileva's method. Consider

$$\frac{dx}{dt} = y \ ; \qquad x(\varepsilon,0) = 1$$

$$\varepsilon \frac{dy}{dt} = x^2 - y^2 ; \qquad y(\varepsilon,0) = 0$$

We begin by analyzing the boundary layer stability properties, which we know to be of key inportance. The ZOBLE for this problem is

$$\frac{dy_b}{d\tau} = g(x_o, y_b) = x_o^2 - y_b^2$$

Thus we are led to consider the stability properties of $dy/dt = x^2 - y^2$ for fixed x The equilibrium points are given by $y = \pm x$ and it is necessary for stability that $\partial g/\partial y = -2y < 0$, ie. $y > 0$. The stable equilibrium points are shown in Figure 1.3, along with their domains of influence. For the specific problem at hand, the initial conditions (1,0) are in the domain of influence of the stable equlibrium point (1,1). We now procede with the formal solution.

The outer problem is

$$\frac{dx^o}{dt} = y^o$$

$$\varepsilon \frac{dy^o}{dt} = x^{o2} - y^{o2}$$

To solve this to first order, substitute (1.10) to get

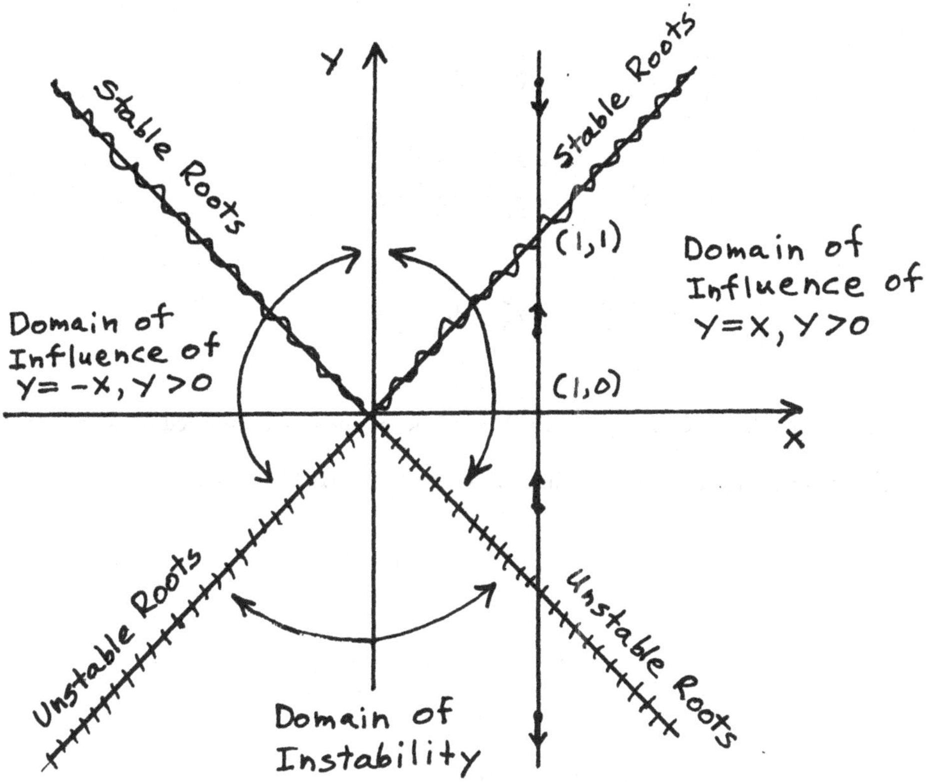


Figure 1.3. Stability domains of $dy/dt = x^2 - y^2$ for paths x = constant.

$$\frac{dx_0^o}{dt} + \frac{dx_1^o}{dt}\varepsilon = y_0^o + y_1^o\varepsilon$$

$$\frac{dy_0^o}{dt}\varepsilon = x_0^{o2} - y_0^{o2} + (2x_0^o x_1^o - 2y_0^o y_1^o)\varepsilon$$

The zeroth order solutions are

$$\frac{dx_o^o}{dt} = y_o^o \;; \qquad 0 = x_o^{o2} - y_o^{o2}$$

$$y_o^o = x_o^o \;; \qquad \frac{dx_o^o}{dt} = x_o^o$$

$$x_o^o = C_o^o e^t; \qquad y_o^o = C_o^o e^t$$

where the stable root $y_o^o = \phi(x_o^o) = x_o^o$ was chosen in accordance with results of the stability analysis. The first order solutions are

$$\frac{dx_1^o}{dt} = y_1^o; \quad \frac{dy_o^o}{dt} = 2x_o^o x_1^o - 2y_o^o y_1^o$$

$$y_1^o = x_1^o - \frac{1}{2}; \quad \frac{dx_1^o}{dt} = x_1^o - \frac{1}{2}$$

$$x_1^o = (C_1^o - \frac{1}{2})e^t + \frac{1}{2}; \quad y_1^o = (C_1^o - \frac{1}{2})e^t$$

The inner problem is

$$\frac{dx^i}{d\tau} = \varepsilon y^i \;; \qquad x^i(\varepsilon,0) = 1$$

$$\frac{dy^i}{d\tau} = x^{i^2} - y^{i^2}; \quad y^i(\varepsilon,0) = 0$$

Substituting (1.15) and retaining terms to first order yields

$$\frac{dx_o^i}{d\tau} + \frac{dx_1^i}{d\tau}\varepsilon = y_o^i\varepsilon; \quad x_o^i(0) + x_1^i(0)\varepsilon = 1$$

$$\frac{dy_o^i}{d\tau} + \frac{dy_1^i}{d\tau}\varepsilon = x_o^{i^2} - y_o^{i^2} + (2x_o^i x_1^i - 2y_o^i y_1^i)\varepsilon \;;$$

$$y_o^i(0) + y_1^i(0)\varepsilon = 0$$

The zeroth order solutions are

$$\frac{dx_o^i}{d\tau} = 0; \qquad x_o^i(0) = 1$$

$$\frac{dy_o^i}{d\tau} = x_o^{i^2} - y_o^{i^2} \;; \qquad y_o^i(0) = 0$$

$$x_o^i = 1 \;; \qquad y_o^i = \tanh\tau$$

The first order solutions are

$$\frac{dx_1^i}{d\tau} = y_o^i \;; \qquad x_1^i(0) = 0$$

$$\frac{dy_1^i}{d\tau} = 2x_o^i x_1^i - 2y_o^i y_1^i \;; \qquad y_1^i(0) = 0$$

$$\frac{dx_1^i}{d\tau} = \tanh\tau \;; \quad \frac{dy_1^i}{d\tau} = 2x_1^i - (2\tanh\tau)y_1^i$$

$$x_1^i = \ell n \cosh \tau \; ; \qquad y_1^i \text{ is not elementary}$$

To sum up, we now have, to first order,

$$x^o = C_o^o e^t + [(C_1^o - \tfrac{1}{2})e^t + \tfrac{1}{2}]\varepsilon$$

$$x^i = 1 + (\ell n \cosh \tau)\varepsilon$$

Expand $x^o$ to first order in t about $t = 0$ to get

$$x^o = C_o^o(1 + t) + [(C_1^o - \tfrac{1}{2})(1 + t) + \tfrac{1}{2}]\varepsilon$$

Expanding $\ell n \cosh \tau$ for large $\tau$ gives

$$\ell n \cosh \tau = \ell n \frac{e^\tau + e^{-\tau}}{2} = \ell n \, e^\tau + \ell n(1 + e^{-2\tau}) - \ell n \, 2$$

$$= \tau - \ell n \, 2 + e^{-2\tau} - \tfrac{1}{2}(e^{-2\tau})^2 + \ldots$$

where terms involving $e^{-2\tau}$ will be smaller than any algebraic term. Thus, to first order,

$$x^i = 1 + (\tfrac{t}{\varepsilon} - \ell n \, 2)\varepsilon$$

Now use the matching relation (1.21) to get $C_1^o$ and $C_o^o$ :

$$\lim_{\substack{\varepsilon\to0\\ t\to0\\ \varepsilon/t\to0}} \{C_o^o(1+t)+[(C_1^o-\tfrac{1}{2})(1+t)+\tfrac{1}{2}]\varepsilon-1-(\tfrac{t}{\varepsilon}-\ell n\ 2)\varepsilon\} = 0$$

$$\lim_{\substack{\varepsilon\to0\\ t\to0\\ \varepsilon/t\to0}} \{C_o^o+C_o^o t+(C_1^o-\tfrac{1}{2})t\varepsilon+(C_1^o-\tfrac{1}{2})\varepsilon+\tfrac{1}{2}\varepsilon-1-t+(\ell n\ 2)\varepsilon\} = 0$$

Equating coefficients of like powers of $t^k\varepsilon^\ell$ gives

$$C_o^o - 1 = 0$$

$$C_o^o - 1 = 0$$

$$C_1^o - \frac{1}{2} + \frac{1}{2} + \ell n\ 2 = 0$$

whence

$$C_o^o = 1 \ ; \quad C_1^o = -\ \ell n\ 2$$

Thus

$$x^o = e^t + [-(\ell n\ 2 + \tfrac{1}{2})e^t + \tfrac{1}{2}]\varepsilon$$

The common part is

$$CP = 1 + t - \varepsilon\ \ell n\ 2$$

The uniform solution (additive composition) to first order in $\varepsilon$ is, according to (1.28),

$$x^a(\varepsilon,t) = e^t+[-(\ell n\, 2+\frac{1}{2})e^t+\frac{1}{2}]\varepsilon+1+(\ell n \cosh \tau)\varepsilon-1-t+\varepsilon\, \ell n\, 2$$

$$x^a(\varepsilon,t) = e^t-t+[-(\ell n\, 2+\frac{1}{2})e^t+\ell n \cosh \frac{t}{\varepsilon}+\frac{1}{2} + \ell n\, 2]\varepsilon$$

For small $\varepsilon$, this provides a useful approximation to the solution of the original problem.

## 2. SINGULARLY PERTURBED NONLINEAR OPTIMAL CONTROL

### 2.1. Problem Formulation and Basic Theorems

We now turn to the nonlinear optimal control problem. Consider the system

$$\begin{aligned} \dot{x} &= f(x,y,u,\varepsilon,t) \\ \varepsilon\dot{y} &= g(x,y,u,\varepsilon,t) \end{aligned} \tag{2.1}$$

on the interval $0 \le t \le T$ subject to initial conditions

$$\begin{aligned} x(\varepsilon,0) &= x_o(\varepsilon) \\ y(\varepsilon,0) &= y_o(\varepsilon) \end{aligned} \tag{2.2}$$

where "$\cdot$" denotes a derivative with respect to t. It is desired to minimize

$$J = \int_0^T \phi(x,y,u,\varepsilon,t)dt \tag{2.3}$$

where T is prescribed. In these equations, $x(\cdot)$ is a "slow" state vector functions with $n_s$ components, $y(\cdot)$ is a "fast" state vector function with $n_f$ components, $u(\cdot)$ is a control vector function with $n_c$ components, and $\varepsilon > 0$ is a parameter. It is assumed that $f(\cdot)$, $g(\cdot)$, $f_x(\cdot)$, $f_y(\cdot)$, $g_x(\cdot)$, and $g_y(\cdot)$ are continuous and that $u(\cdot)$ is piecewise continuous and unconstrained for $0 \le t \le T$, where subscripts denote partial differentiation.

Rewrite (2.1) as

$$\begin{aligned} \dot{x} &= f(x,y,u,\varepsilon,t) \\ \dot{y} &= \frac{1}{\varepsilon} g(x,y,u,\varepsilon,t) \end{aligned} \tag{2.4}$$

Let $\lambda_x(\cdot)$ and $\lambda_y(\cdot)$ be any nonzero vector functions of dimensions $n_s$ and $n_f$, respectively, such that their components satisfy the following linear system of equations. (The usual prime notation for the transpose of a matrix would prove to be cumbersone in the sequal, and we therefore omitt it; it should be obvious from the context whether a matrix or its transpose is implied.)

$$\begin{aligned} \dot{\lambda}_x &= -\phi_x\lambda_o - f_x\lambda_x - \frac{1}{\varepsilon} g_x\lambda_y \\ \dot{\lambda}_y &= -\phi_y\lambda_o - f_y\lambda_x - \frac{1}{\varepsilon} g_y\lambda_y \end{aligned} \tag{2.5}$$

Define the scalar function $H'(\cdot)$ by

$$H'(\lambda_o,\lambda_x,\lambda_y,x,y,u,\varepsilon,t) = \phi\lambda_o + f\lambda_x + \frac{1}{\varepsilon} g\lambda_y \tag{2.6}$$

Introducing the transformation

$$\lambda_x = \lambda \ ; \quad \lambda_y = \varepsilon\mu \tag{2.7}$$

into (2.5) and (2.6) results in

$$\begin{aligned} \dot{\lambda} &= -\phi_x\lambda_o - f_x\lambda - g_x\mu \\ \varepsilon\dot{\mu} &= -\phi_y\lambda_o - f_y\lambda - g_y\mu \end{aligned} \tag{2.8}$$

and

$$H(\lambda_o,\lambda,\mu,x,y,u,\varepsilon,t) = \phi\lambda_o + f\lambda + g\mu \tag{2.9}$$

respectively. Applying the well-known Pontryagin Maximum Principle ([13]-[15]) to this problem then gives the following necessary conditions for optimal control.

Theorem 2.1. (Maximum Principle for Singularly Perturbed Systems.) If the control $u(\cdot)$ minimizes (2.3) and, along with $x(\cdot)$ and $y(\cdot)$, satisfy (2.1) and (2.2) then there exist nonzero functions $\lambda(\cdot)$ and $\mu(\cdot)$ whose components satisfy (2.8) such that

(a) $H_u(\lambda_o,\lambda(\varepsilon,t),\mu(\varepsilon,t),x(\varepsilon,t),y(\varepsilon,t),u,\varepsilon,t) = 0$

(b) $\lambda_o = \text{constant} \le 0$

(c) $\lambda(\varepsilon,T) = 0$ and $\mu(\varepsilon,T) = 0$.

Thus the problem reduces to one of solving the following two point boundary value problem (from now on functional dependence will be omitted when it does not result in a lack of clarity)

$$\begin{aligned}
\dot{x} &= f \\
\varepsilon\dot{y} &= g \\
\dot{\lambda} &= -H_x \\
\varepsilon\dot{\mu} &= -H_y \\
0 &= H_u \\
x(\varepsilon,0) &= x_o(\varepsilon); \quad y(\varepsilon,0) = y_o(\varepsilon) \\
\lambda(\varepsilon,T) &= 0 \; ; \quad \mu(\varepsilon,T) = 0
\end{aligned} \tag{2.11}$$

where H(•) is given by (2.9). We note that in this formulation of the problem the adjoint variables associated with the slow state variables are themselves slow and the adjoint variables associated with the fast state variables are themselves fast. If it is assumed that a unique optimal control exists, then (2.11) has a unique solution. We shall in fact assume that $H_{uu}$ is negative definite in the subsequent discussion.

Several generalizations of this basic problem are possible, for example to the cases in which: (i) the final time T is not prescribed, (ii) the cost J depends on the values of the state variables at the final time, (iii) the control function u(•) is bounded, and (iv) the system is transfered from a manifold in state space of dimension equal to or less than the total system dimension $n_s + n_f$ to another such manifold. For our purposes, these generalizations (except for the one introducing bounded control) would only add algebraic complexity to the analysis.

Asymptotic analysis of systems of singularly perturbed differential equations that arise in nonlinear optimal control have been undertaken in [16] - [20], and results similar to the results of Tihonov and Vasileva for the system (1.1) have been obtained for the system (2.11). As before, we call the system with $\varepsilon = 0$ and the boundary conditions on the fast variables omitted the reduced problem:

$$\begin{aligned} \dot{x}_r &= f_r \\ 0 &= g_r \\ \dot{\lambda}_r &= -H_{x_r} \\ 0 &= -H_{y_r} \\ 0 &= H_{u_r} \\ x_r(0) &= x_o(0); \; \lambda_r(T) = 0 \end{aligned} \tag{2.12}$$

where, for example, $f_r$ denotes $f(x_r, y_r, u_r, 0, t)$; the reduced state, adjoint, and control variables are of course functions only of t. Because of the two point boundary value nature of this problem, there will generally be two boundary layer problems associated with (2.11), called herein the initial and the terminal.

Based on the discussion in Section 1.1 we anticipate that the following types of assumptions will be required to insure the proper asymptotic behavior of the solutions of (2.11): (A) additional smoothness of the system functions, (B) existence of a solution to the reduced problem (2.12),

(C) satisfaction of an eigenvalue criterion that insures local boundary layer stability, (D) satisfaction of a requirement that the stable boundary layer solutions have enough independent parameters to satisfy all relevant boundary conditions, and (E) assurance that the boundary conditions are in the domain of influence of the reduced solution evaluated at the boundaries. Under these assumptions it may be expected that: (i) a solution to the problem (2.11) exists, (ii) the solution to the full problem tends to the solution of the reduced problems as $\varepsilon \to 0$ everywhere for the slow variables x and $\lambda$ and everywhere except at the boundaries for the fast variables y and $\mu$, and (iii) asymptotically valid expansions of the solution to the full problem (2.11) exist up to an order in $\varepsilon$ related to the degree of system smoothness. Before making these statements precise, we will briefly review the local stability properties of the initial zero order boundary layer equation (ZOBLE). These properties have been derived and discussed, for example, in [21] and only the results will be presented here.

To obtain the initial ZOBLE, as before we substitute

$$\tau = \frac{t}{\varepsilon} \tag{2.13}$$

in (2.11) and set $\varepsilon = 0$ to get

$$\left.\begin{aligned} \frac{dy_b}{d\tau} &= g_b \\ \frac{d\mu_b}{d\tau} &= -H_{yb} \\ 0 &= H_{u_b} \\ y_b(0) &= y_o(0) \end{aligned}\right\} \tag{2.14}$$

where, for example, $H_{y_b}$ denotes $\partial H(\lambda_o, \lambda_r(0), \mu_b, x_o(0), y_b, u_b, 0, 0)/\partial y_b$. As shown in [21], the perturbation equations obtained by linearizing (2.14) about the outer solution evaluated at $t = 0$, which is an equilibrium point of (2.14), has a constant $2n_f \times 2n_f$ coefficient matrix of the form

$$G_{r_o} = \begin{bmatrix} A_{r_o} & B_{r_o} \\ C_{r_o} & -A'_{r_o} \end{bmatrix} \tag{2.15}$$

where

$$\left.\begin{aligned} A_{r_o} &= g_{y_{r_o}} - g_{u_{r_o}} H^{-1}_{uu_{r_o}} H_{uy_{r_o}} \\ B_{r_o} &= -g_{u_{r_o}} H^{-1}_{uu_{r_o}} g'_{u_{r_o}} = B'_{r_o} \\ C_{r_o} &= -H_{yy_{r_o}} + H_{yu_{r_o}} H^{-1}_{uu_{r_o}} H_{uy_{r_o}} = C'_{r_o} \end{aligned}\right\} \tag{2.16}$$

and where subscript $r_o$ indicates that these matrices are to be evaluated on the reduced solution evaluated at $t = 0$.

Since there are $n_f$ boundary conditions specified for the $2n_f$ equations (2.14), we must have at least $n_f$ stable modes, i.e. the matrix $G_{r_o}$ must have at least $n_f$ eigenvalues with negative real parts. However, a well known property of a matrix with the structure of (2.15) is that if s is an eigenvalue than so is -s and therefore we will have the property we want if and only if there are no eigenvalues with zero real parts. Since a similar result holds for the terminal boundary layer, the local eigenvalue criterion for boundary layer stability of (2.11) can be stated as

"There are no eigenvalues of $G_r(t)$
with zero real parts on $0 \le t \le T$" (2.17)

Before stating the basic theorem giving the asymptotic properties of the solution of (2.11), one more matrix must be introduced. Let P be a nonsignular $2n_f \times 2n_f$ matrix such that

$$P^{-1}GP = \begin{bmatrix} D_1 & 0 \\ 0 & D_2 \end{bmatrix} \tag{2.18}$$

where $D_1$ has only eigenvalues with negative real part and $D_2$ has only eigenvalues with positive real part. If the eigenvalue criterion on G just stated is satisfied on $0 \le t \le T$, then such a matrix P exists on $0 \le t \le T$. Partition P into the form

$$P = \begin{bmatrix} P_{11} & P_{12} \\ P_{21} & P_{22} \end{bmatrix} \tag{2.19}$$

where all the $P_{ij}$ are $n_f \times n_f$ matrices. We are now ready to state the following result, which is based on [18] and [19]; the formal hypotheses correspond to the qualitative ones made earlier.

Theorem 2.2. Consider the system (2.11) and suppose that the following are satisfied: (A) there exists an $\varepsilon_o > 0$ such that f and g are K + 2 times and H is K + 3 times continuously differentiable with respect to $x,y,u,\varepsilon$ and t and $x_o$ and $y_o$ are K + 2 times continuously differentiable with respect to $\varepsilon$, for all $0 \le t \le T$ and $0 \le \varepsilon \le \varepsilon_o$, in a neighborhood of the reduced solution; (B) the reduced system (2.12) has a continuous solution on $0 \le t \le T$; (C) the matrix G defined by (2.15) satisfies the eigenvalue criterion (2.17); (D) $P_{11}(0)$ and $P_{22}(T)$ as defined by (2.19) are nonsingular; and (E) the quantities $|y_o(0) - y_r(0)|$ and $|\mu_r(0)|$ are sufficiently small to insure that the initial and terminal boundary conditions are in the domains of influence of the reduced solution evaluated at t = 0 and t = T, respectively. Then, for $0 \le \varepsilon \le \varepsilon_o$: (i) the full system (2.11) has a unique solution; (ii) the solution of of (2.11) for x and $\lambda$ tends to the solution of (2.12) for $x_r$ and $\lambda_r$ uniformly on $0 \le t \le T$ and the solution of (2.11) for y and $\mu$ tends to the solution of (2.12) for $y_r$ and $\mu_r$ uniformly on any closed subinterval of $0 < t < T$, as $\varepsilon$ tends to zero; (iii) there exist solutions to the initial and terminal ZOBLES which are asymptotically stable with respect to the reduced solution and which satisfy all imposed boundary conditions; and

(iv) the outer, initial boundary layer, and terminal boundary layer systems associated with (2.11) all possess asymptotically valied expansions in $\varepsilon$ up to order K such that, when suitably combined, they give an asymptotically valid expansion of the solution of (2.11) up to order K in $\varepsilon$.

It should be remarked that in the course of the proofs of this result given in [18] and [19] explicit formulas are derived for the required expansions.

The eigenvalue criterion used here, (2.17), is actually somewhat weaker than generally adapted (it is often required that the matrix $A_r$ in (2.16) be stable). Nevertheless, there are indications that it can be further weakend. For example, [22] indicates that eigenvalues with zero real parts maybe allowed, provided a form of conditional stability is satisfied. From an applications standpoint, successful numerical solutions of problems with unstable boundary layers have been achieved in fluid mechanics, as for example in [23] and [24].

The special case of linear state equations and quadratic cost function has been extensively analyzed (see [25]-[27] for reviews of this work). Analysis of the linear-quadratic problem, as given for example in Chapter 3 of [2], is useful as an aid to understanding the nonlinear problem we are considering here because it shows very clearly the role played by each of the hypotheses of Theorem 2.2. The case where system nonlinearity is alowed in x (but not in y or u) is also relatively simple and has been treated by many authors; see for example [27], [28], and, especially, [29].

In practice, as has been remarked earlier, the only condition of the Theorem which is generally useful is the eigenvalue criterion (2.17). This condition is relatively easy to check and gives valuable information regarding the behavior of the ZOBLES. It's application in a complex nonlinear flight dynamics problem is described in [30] and [31].

In the case where f and g are scalar functions, the eigenvalue criterion takes on an especially simple form, namely that

$$H_{yy_r} g_{u_r}^2 - 2H_{yu_r} g_{y_r} g_{u_r} + H_{uu_r} g_{y_r}^2 < 0 \tag{2.20}$$

must be satisfied at t = 0 and at t = T. This result was first derived in [32] where it was observed that (2.20) is just the strenghtened form of the Legendre Clebzch condition of the calculus of variations for the reduced problem. Equation (2.20) is also derived and discussed in [2].

## 2.2. Alternative Procedure for the Reduced Problem

In some applications, the reduced solution is a sufficiently good approximation. In this case, the natural question arises as to when it is possible to set $\varepsilon = 0$ before applying the Maximum Principle instead of after. The former procedure is attractive because it involves less algebraic manipulation . The following, taken from section 5.4 of [2], anwers this question.

Theorem 2.3. In addition to the assumptions of Section 2.1, suppose that the matrix $[g_y g_u]$ has maximum rank (i.e. rank $n_f$) evaluated along the reduced solution. Then the reduced problem is the same as the problem obtained by the alternative proc4dure of setting $\varepsilon = 0$ in the state equations and applying the necessary conditions for optimal control to the result.

Proof. The proof follows Section 3.6 of [14]. Since $[g_y g_u]$ has rank $n_f$, by the implicit function theorem $0 = g(x,y,u,\varepsilon,t)$ can be solved for $n_f$ of the components of (y,u), say p, in terms of the remaining $n_c$ components, say q. Thus

$$p = \psi(q,x)$$

Further, $g_p$ is nonsingular so that

$$\psi_x = -g_p'^{-1} g_x'$$

where $g'(x,p,q,\varepsilon,t) = g(x,y,u,\varepsilon,t)$. Now consider (2 ) with $\varepsilon = 0$ and (y,u) replaced by (p,q) and let, for example, $f(x,y\ \ ,\varepsilon,t) = f'(x,p,q,\varepsilon,t)$:

$$\dot{x} = f'(x,p,q,0,t)$$

$$0 = g'(x,p,q,0,t)$$

The adjoint variable $\lambda$ is a solution of

$$\dot{\lambda} = -(\phi_x' + \phi_p'\psi_x)\lambda_o - (f_x' + f_p'\psi_x)\lambda$$

i.e.

$$\dot{\lambda} = -\phi_x'\lambda_o - f_x'\lambda + (\phi_o'\lambda_o + f_p'\lambda)g_p'^{-1} g_x'$$

The H function is

$$H = \phi'\lambda_o + f'\lambda$$

Since this must be maximized subject to $g' = 0$, we introduce the ordinary Lagrange multiplier $\nu$ and regard $(p,q)$ as the control vector. We form

$$H' = \phi'\lambda_o + f'\lambda + g'\nu$$

For the maximum of $H'$ (i.e. a maximum of H subject to $g' = 0$) it is necessary that

$$H'_p = \phi'_p\lambda_o + f'_p\lambda + g'_p\nu = 0$$

$$H'_q = \phi'_q\lambda_o + f'_q\lambda + g'_q\nu = 0$$

From the first of these

$$\nu = -g'^{-1}_p(\phi'_p\lambda_o + f'_p\lambda)$$

so that

$$\dot{\lambda} = -\phi'_x\lambda_o - f'_x\lambda - g'_x\nu$$

Therefore, the alternative procedure results in the problem

$$\dot{x} = f'$$

$$0 = g'$$

$$\dot{\lambda} = -\phi_x'\lambda_o - f_x'\lambda - g_x'\nu$$

$$0 = \phi_p'\lambda_o + f_p'\lambda + g_p'\nu$$

$$0 = \phi_q'\lambda_o + f_q'\lambda + g_q'\nu$$

Now consider the reduced problem (2.12). Changing variables (y,u) to (p,q) results in

$$\dot{x} = f'$$

$$0 = g'$$

$$\dot{\lambda} = -\phi_x'\lambda_o - f_x'\lambda - g_x'\mu$$

$$0 = \phi_p'\lambda_o + f_p'\lambda + g_p'\mu$$

$$0 = \phi_q'\lambda_o + f_q'\lambda + g_q'\mu$$

Thus, the two procedures result in the same problem. The variational multiplier (adjoint variable) $\mu$ has become an ordinary multiplier $\nu$ in the reduced problem.

The significance of the hypothesis of the theorem is clear. If it is satisfied, $n_f$ of the components of (y,u) may be eliminated from the problem. Alternatively, (y,u) may be regarded as the control vector which results in a problem with state dependent control constraints. Theorem

2.3 generalizes Hadlock's result [16] which gives the nonsingularity of $g_y$ as a sufficient condition for the equivalence of procedures.

## 3. SOLUTION BY MATCHED ASYMPTOTIC EXPANSIONS

### 3.1. Analysis

Several authors have developed asymptotic methods for the analysis of nonlinear optimal control problems based generally on the method introduced by Vasileva alluded to in Section 1; [18] and [19] particularly should be mentioned in this regard. An early work along these lines was [33], but the development there was strictly formal and focused on the special problems associated with bounded control.

An interesting procedure has been proposed in [34] and [25] and will be briefly described. If the state equations can be cast in the form

$$\begin{aligned} &\dot{x} = f \quad ; \quad \dim x = n_s \quad ; \quad x(0) = x_o \\ &\varepsilon^i \dot{y}_i = g_i ; \; i = 1,2,\ldots,n_f ; \quad y_i(0) = y_{io} \end{aligned} \tag{3.1}$$

Then there will be a sequence of initial and terminal boundary layer problems, each of which is a second order system subject to one boundary condition. If, additionally, the system is autonomous and the final time T is not prescribed, the function H defined in (2.9) is identically zero. This condition can be used to furnish the second constant of integration for the boundary layer systems with the result that two point boundary value problems, which generally require iterative solution, are avoided in the boundary layer problems.

In this section, we will apply the method of matched asymptotic exlansion to the nonlinear optimal control problem. Such an approach has been developed for a specific problem in aircraft flight mechanics in [2] and [30] and was later adapted to obtain first order corrections for other similar problems (see [35] and [36]). Our goal is to now develope this method for the general problem (2.11). The analysis will follow the same pattern as in Section 1.2: (i) formulate the outer and boundary layer problems, (ii) obtain asymptotic solutions to these problems, (iii) match these solutions to obtain all constants of integration, and (iv) form the additive composite to obtain a uniformly valied asymptotic representation of the solution to the original problem. The reasons for using the MAE method are the same as those stated in Section 1.2. We assume that a unique, unbounded optimal control exists and that all hypotheses of Theorem 2.2 hold. An analysis similar to that which follows may be found in [2] for the special case of the linear/quadratic problem.

The outer system is simply (2.11) without the boundary conditions; denoting the outer solution by, for example, $x^o(\varepsilon,t)$, we have

$$\begin{aligned} \dot{x}^o &= f^o \\ \varepsilon\dot{y}^o &= g^o \\ \dot{\lambda}^o &= -H_x^o \\ \varepsilon\dot{\mu}^o &= -H_y^o \\ 0 &= H_u^o \end{aligned} \tag{3.2}$$

where, for example, $H_x^o = \partial H(\lambda_o,\lambda^o,\mu^o,x^o,y^o,u^o,\varepsilon,t)/\partial x^o$.

The initial boundary layer system of equations is obtained as before by introducing the stretching transformation

$$\tau = \frac{t}{\varepsilon} \tag{3.3}$$

into (2.11); the result is, denoting the solution by, for example, $x^{i1}(\varepsilon,\tau)$,

$$\begin{aligned}
\frac{dx^{i1}}{d\tau} &= \varepsilon f^{i1} \quad ; \quad x^{i1}(\varepsilon,0) = x_o(\varepsilon) \\
\frac{dy^{i1}}{d\tau} &= g^{i1} \quad ; \quad y^{i1}(\varepsilon,0) = y_o(\varepsilon) \\
\frac{d\lambda^{i1}}{d\tau} &= -\varepsilon H_x^{i1} \\
\frac{d\mu^{i1}}{d\tau} &= -H_y^{i1} \\
0 &= H_u^{i1}
\end{aligned} \tag{3.4}$$

where, for example, $H_x^{i1} = \partial H(\lambda_o,\lambda^{i1},\mu^{i1},x^{i1},y^{i1},u^{i1},\varepsilon,\varepsilon\tau)/\partial x^{i1}$

Similarly, the terminal boundary layer system is obtained by stretching the time-to-go by $\varepsilon$,

$$\sigma = \frac{T-t}{\varepsilon} \tag{3.5}$$

The result is, denoting the solution by, for example, $x^{i2}(\varepsilon,\sigma)$,

$$\frac{dx^{i2}}{d\sigma} = -\varepsilon f^{i2}$$

$$\frac{dy^{i2}}{d\sigma^{i2}} = -g^{i2}$$

$$\frac{d\lambda^{i2}}{d\sigma} = \varepsilon H_x^{i2} \ ; \quad \lambda^{i2}(\varepsilon,0) = 0 \tag{3.6}$$

$$\frac{d\mu^{i2}}{d\sigma} = H_y^{i2} \ ; \quad \mu^{i2}(\varepsilon,0) = 0$$

$$0 = H_u^{i2}$$

where, for example, $H_x^{i2} = \partial H(\lambda,\lambda^{i2},\mu^{i2},x^{i2},y^{i2},u^{i2},\varepsilon,T-\varepsilon\sigma)/\partial x^{i2}$

To solve (3.2) we express all dependent variables $(x^o,y^o,\lambda^o,\mu^o,u^o)$ in asymptotic power series, for example,

$$x^o(\varepsilon,t) \sim \sum_{j=0}^{k} x_j^o(t)\varepsilon^j; \quad k \le K \tag{3.7}$$

This leads to the sequence of $2h_s$ dimensional problems

$$
\begin{aligned}
\dot{x}_o^o &= f_o^o & \dot{x}_j^o &= f_j^o \\
0 &= g_o^o & \dot{y}_{j-1}^o &= g_j^o \\
\dot{\lambda}_o^o &= -H_{x_o}^o & \dot{\lambda}_j^o &= -H_{x_j}^o \\
0 &= -H_{y_o}^o & \dot{\mu}_{j-1}^o &= -H_{y_j}^o \\
0 &= H_{u_o}^o & 0 &= H_{u_j}^o \\
& & j &= 1,\dots,k
\end{aligned}
\tag{3.8}
$$

each of which is linear except for the first. In (3.8), the functions such as $f_j^o$ are obtained by expansion about $\varepsilon = 0$, that is from

$$
\begin{aligned}
f^o = f(x^o,y^o,u^o,\varepsilon,t) &= f(x_o^o,y_o^o,u_o^o,0,t) \\
&+ (f_{x_o}^o x_1^o + f_{y_o}^o y_1^o + f_{u_o}^o u_1^o + f_{\varepsilon_o}^o)\varepsilon \\
&+ \dots = f_o^o + f_1^o\,\varepsilon + \dots
\end{aligned}
\tag{3.9}
$$

where, for example, $f_{x_o}^o = \partial f(x_o^o,y_o^o,u_o^o,0,t)/\partial x_o^o$

Next, to solve (3.4) we put, for example,

$$x^{il}(\varepsilon,\tau) \sim \sum_{j=0}^{k} x_j^{il}(\tau)\varepsilon^j; \quad k \le K \tag{3.10}$$

in (3.4) to get a sequence of $2n_f$ dimensional problems

$$\begin{aligned}
&\frac{dx_o^{il}}{d\tau} = 0 \quad ; \quad x_o^{il}(0) = x_{oo} \\
&\frac{dy_o^{il}}{d\tau} = g_o^{il} \quad ; \quad y_o^{il}(0) = y_{oo} \\
&\frac{d\lambda_o^{il}}{d\tau} = 0 \\
&\frac{d\mu_o^{il}}{d\tau} = -H_{y_o}^{il} \\
&0 = H_{u_o}^{il} \\
&\frac{dx_j^{il}}{d\tau} = f_{j-1}^{il} \quad ; \quad x_j^{il}(0) = x_{oj} \\
&\frac{dy_j^{il}}{d\tau} = g_j^{il} \quad ; \quad y_j^{il}(0) = y_{oj} \\
&\frac{d\lambda_j^{il}}{d\tau} = -H_{x_{j-1}}^{il} \\
&\frac{d\mu_j^{il}}{d\tau} = -H_{y_j}^{il} \\
&0 = H_{u_j}^{il} \quad ; \quad j = 1,\dots,k
\end{aligned} \tag{3.11}$$

where the functions such as $f_j^{i1}$ are obtained by expansion about $\varepsilon = 0$ as before, and $x_{oj}$ and $y_{oj}$ are given by

$$x_o(\varepsilon) = \sum_{j=1}^{k} x_{oj}\varepsilon^j \; ; \quad y_o(\varepsilon) = \sum_{j=1}^{k} y_{oj}\varepsilon^j \tag{3.12}$$

As before, the first in the sequence of problems (3.11) is nonlinear and the rest are linear.

Finally, the terminal boundary layer system (3.6) is solved asymptotically by setting, for example

$$x^{i2}(\varepsilon,\sigma) \sim \sum_{j=0}^{k} x_j^{i2}(\sigma)\varepsilon^j ; \quad k \le K \tag{3.13}$$

to get the sequence of $2n_f$ dimensional problems

$$\begin{aligned}
\frac{dx_o^{i2}}{d\sigma} &= 0 \\
\frac{dy_o^{i2}}{d\sigma} &= -g_o^{i2} \\
\frac{d\lambda_o^{i2}}{d\sigma} &= 0 \qquad ; \quad \lambda_o^{i2}(0) = 0 \\
\frac{d\mu_o^{i2}}{d\sigma} &= H_{y_o}^{i2} \qquad ; \quad \mu_o^{i2}(0) = 0 \\
0 &= H_{u_o}^{i2} \\
\frac{dx_j^{i2}}{d\sigma} &= -f_{j-1}^{i2}
\end{aligned} \tag{3.14}$$

$$\frac{dy_j^{i2}}{d\sigma} = -g_j^{i2}$$

$$\frac{d\lambda_j^{i2}}{d\sigma} = H_{x_{j-1}}^{i2} \qquad ; \lambda_j^{i2}(0) = 0$$

$$\frac{d\mu_j^{i2}}{d\sigma} = H_{y_j}^{i2} \qquad ; \mu_j^{i2}(0) = 0 \qquad (3.14)\ \text{cont.}$$

$$0 = H_{u_j}^{i2}$$

$$j = 1,\ldots,k$$

only the first of which is nonlinear.

Usually only the first two terms of these expansions are used in practice. This is due not only to the algebraic complexity of higher order terms but also to the fact that if a satisfactory solution is not obtained after two terms then higher order terms are not likely to improve the situation. Therefore our investigation will be limited to the first two terms. Consider first the leading problem of (3.8), (3.11) and (3.14), i.e. the reduced system with the boundary conditions removed and the initial and terminal ZOBLES. Solution of the first of the problems (3.8) will contain $2n_s$ as yet unknown constants of integration, say $x_o^o(0)$ and $\lambda_o^o(T)$. For the leading problem of (3.11) we have $x_o^{i1}(\tau) = x_{oo}$ and $\lambda_o^{i1}(\tau) = \lambda_o^{i1}(0)$, the later an unknown vector constant. The remaining $2n_s$ equations have only $n_s$ boundary conditions. We use the other $n_s$ boundary conditions, say $\mu_o^{i1}(0)$, to sppress the "unstable modes". We know that we will have precisely the right number of free boundary conditions to do this and that the "stable modes" will be able to satisfy all the prespecified boundary conditions, $y_o^{io}(0) = y_{oo}$. Similarly, for the leading problem of (3.14), $x_o^{i2}(\sigma) = x_o^{i2}(0)$ and $\lambda_o^{i2}(\sigma) = 0$, the former an unknown constant. We use the $n_f$ free constants $y_o^{i2}(0)$ to suppress the "unstable modes", leaving just enough "stable modes" to satisfy the prespecified boundary conditions, $\mu_o^{i2}(0) = 0$.

We now can match the slow variables, x and $\lambda$, at t = 0 and t = T to zero order to obtain the unknown constants of integraion in the outer solution. At t = 0, the matching rule (1.20) applied to x and $\lambda$ to zero order gives simply

$$\begin{aligned} x_o^o(0) &= x_{oo} \\ \lambda_o^o(0) &= \lambda_o^{i1}(0) \end{aligned} \tag{3.15}$$

and at t = T a similar rule implies

$$\begin{aligned} x_o^o(T) &= x_o^{i2}(0) \\ \lambda_o^o(T) &= 0 \end{aligned} \tag{3.16}$$

The zero order problem is now fully determined. First, solve

$$\left.\begin{aligned} &\dot{x}_o^o = f_o^o \quad ; \quad \dot{\lambda}_o^o = -H_{x_o}^o \\ &\text{subject to} \\ &0 = g_o^o \; ; \; 0 = H_{y_o}^o \; ; \; 0 = H_{u_o}^o \\ &x_o^o(0) = x_{oo} \quad ; \quad \lambda_o^o(T) = 0 \end{aligned}\right\} \tag{3.17}$$

Next, solve

$$\left.\begin{aligned}
&\frac{dy_o^{i1}}{d\tau} = g_o^{i1} \quad ; \quad \frac{d\mu_o^{i1}}{d\tau} = -H_{y_o}^{i1} \\
&\text{subject to} \\
&x_o^{i1} = x_{oo} \; ; \; \lambda_o^{i1} = \lambda_o^o(0); \quad 0 = H_{u_o}^{i1} \\
&y_o^{i1}(0) = y_{oo}; \quad \mu_o^{i1}(0) \text{ selected to suppress instability}
\end{aligned}\right\} \tag{3.18}$$

And finally,

$$\left.\begin{aligned}
&\frac{dy_o^{i2}}{d\sigma} = -g_o^{i2} \quad ; \quad \frac{d\mu_o^{i2}}{d\sigma} = H_{y_o}^{i2} \\
&\text{subject to} \\
&x_o^{i2} = x_o^o(T) \; ; \; \lambda_o^{i2} = 0; \quad 0 = H_{u_o}^{i2} \\
&y_o^{i2}(0) \text{ selected to suppress instability}; \quad \mu_o^{i2}(0) = 0
\end{aligned}\right\} \tag{3.19}$$

The problem (3.17) is a $2n_s$ dimensional two-point boundary value problem (2PBVP) on a finite interval T. Problem (3.18) and (3.19) are essentially $2n_f$ dimensional 2PBVPs on an infinite interval but in practice they would be solved on time intervals $\tau^*$ and $\sigma^*$, respectively, where $\tau^*$ and $\sigma^*$ are sufficiently large such that the transients have become negligably small; $\tau^*$ and $\sigma^*$ of course depend on $\varepsilon$. Thus, in effect, we have approximated the solution to a $2(n_s + n_f)$ dimensional 2PBVP by the solutions to one $2n_s$ and two $2n_f$ dimensional problems.

For forming additive composite solutions which are valid everywhere on $0 \le t \le T$ for all variables, in the fashion of (1.29), the common parts will be needed. Since there are two boundary layers, each variable will have two common parts. For the zero order, these common parts are simply the values of the reduced solution variables evaluated at the boundaries:

$$
\begin{aligned}
CP^{i1}_{x_o} &= x_{oo} & CP^{i2}_{x_o} &= x^o_o(T) \\
CP^{i1}_{y_o} &= y^o_o(0) & CP^{i2}_{y_o} &= y^o_o(T) \\
CP^{i1}_{\lambda_o} &= \lambda^o_o(0) & CP^{i2}_{\lambda_o} &= 0 \\
CP^{i1}_{\mu_o} &= \mu^o_o(0) & CP^{i2}_{\mu_o} &= \mu^o_o(T) \\
CP^{i1}_{u_o} &= u^o_o(0) & CP^{i2}_{u_o} &= u^o_o(T)
\end{aligned}
\qquad (3.20)
$$

The additive composite solution for each variable is formed according to, for example,

$$
\begin{aligned}
x^a_o(\varepsilon,t) = x^o_o(t) + x^{i1}_o\left(\frac{t}{\varepsilon}\right) + x^{i2}_o\left(\frac{T-t}{\varepsilon}\right) \\
- CP^{i1}_{x_o}(\varepsilon,t) - CP^{i2}_{x_o}(\varepsilon,t)
\end{aligned}
\qquad (3.21)
$$

The result is

$$
\begin{aligned}
x_o^a(\varepsilon,t) &= x_o^o(t) \\
y_o^a(\varepsilon,t) &= y_o^o(t) + y_o^{i1}(\tfrac{t}{\varepsilon}) + y_o^{i2}(\tfrac{T-t}{\varepsilon}) - y_o^o(0) - y_o^o(T) \\
\lambda_o^a(\varepsilon,t) &= \lambda_o^o(t) \\
\mu_o^a(\varepsilon,t) &= \mu_o^o(t) + \mu_o^{i1}(\tfrac{t}{\varepsilon}) + \mu_o^{i2}(\tfrac{T-t}{\varepsilon}) - \mu_o^o(0) - \mu_o^o(T) \\
u_o^a(\varepsilon,t) &= u_o^o(t) + u_o^{i1}(\tfrac{t}{\varepsilon}) + u_o^{i2}(\tfrac{T-t}{\varepsilon}) - u_o^o(0) - u_o^o(T)
\end{aligned}
\qquad (3.22)
$$

We note that the additive composition to the zero order for the slow variables $x$ and $\lambda$ is just the reduced solution, which is independent of $\varepsilon$. For the fast variables and the control, the composite solution consists of, for example for $y$, the reduced solution $y_o^o(t)$ augmented by boundary layer corrections due to the initial layer, $[y_o^{i1}(\frac{t}{\varepsilon}) - y_o^o(0)]$, and to the terminal layer, $[y_o^{i2}(\frac{T-t}{\varepsilon}) - y_o^o(T)]$.

We now use (3.22) to evaluate $y_o^a(\varepsilon,0)$ and $\mu_o^a(\varepsilon,0)$ as a check to see if their boundary conditions are satisfied. The result is

$$
\begin{aligned}
y_o^a(\varepsilon,0) &= y_{oo} + [y_o^{i2}(\tfrac{T}{\varepsilon}) - y_o^o(T)] \\
\mu_o^a(\varepsilon,T) &= [\mu_o^{i1}(\tfrac{T}{\varepsilon}) - \mu_o^o(0)]
\end{aligned}
\qquad (3.23)
$$

Because of boundary layer stability,

$$\lim_{T/\varepsilon\to\infty} y_o^{i2}(\frac{T}{\varepsilon}) = y_o^o(T)$$
$$\lim_{T/\varepsilon\to\infty} \mu_o^{i1}(\frac{T}{\varepsilon}) = \mu_o^o(0) \tag{3.24}$$

Thus the bracheted terms in (3.23) will be asymptotically negligible although not in general numerically zero and the boundary conditions on y and $\mu$ will not be met exactly; the larger the value of $T/\varepsilon$ the smaller will be the error. This error in boundary conditions is a consequence of "each boundary layer not knowing of the other's existence". In the practical case in which the boundary layer integraions are performed on finite intervals $\tau^*$ and $\sigma^*$, the boundary conditions will be met exactly provided that $\tau^* < T/\varepsilon$ and $\sigma^* < T/\varepsilon$, i.e. provided that each boundary layer has "died out" before the other boundary has been reached. It is logical to make this condition a requirement, since for the asymptotic solution to give a good numerical approximation requires that the boundary layer motion be relatively insignificant compared to the outer motion, or to put it another way, a "strong seperation between the slow and fast variables" and a relatively long time interval. Thus we impose the requirements

$$T - \varepsilon\tau^*(\varepsilon) > 0 \ ; \quad T - \varepsilon\sigma^*(\varepsilon) > 0 \tag{3.25}$$

The larger the values of $T - \varepsilon\tau^*(\varepsilon)$ and $T - \varepsilon\sigma^*(\varepsilon)$, the better will the asymptotic solution numerically approximate the exact solution. We do, however, allow the boundary layes to "overlap", that is it is possible that $\varepsilon\tau^* + \varepsilon\sigma^* > T$.

We remark that the application of the formal MAE procedure is essentially trivial for the zero order. We already knew that the zero order outer problem is just the reduced problem and formation of the common parts and the additive composite solutions can be done by inspection. However, for higher order terms the appropriate boundary conditions and common parts are not so obvious and the procedure of the method must be strictly applied. The following relations, obtained from (2.9), will be useful in deriving the higher order terms:

$$
\begin{aligned}
H_{x\lambda} &= f_x \qquad & H_{x\mu} &= g_x \\
H_{y\lambda} &= f_y \qquad & H_{y\mu} &= g_y \\
H_{u\lambda} &= f_u \qquad & H_{u\mu} &= g_u
\end{aligned}
\tag{3.26}
$$

The first order problems will now be derived and discussed. The first order outer problem comes from (3.8) with $j = 1$; after using (3.26) and rearanging, there results

$$
\begin{aligned}
\dot{x}^o_1 &= f^o_{x_o} x^o_1 + f^o_{y_o} y^o_1 + f^o_{u_o} u^o_1 + f^o_{\varepsilon_o} \\
\dot{\lambda}^o_1 &= -H^o_{xx_o} x^o_1 - f^o_{x_o} \lambda^o_1 - H^o_{xy_o} y^o_1 - g^o_{x_o} \mu^o_1 - H^o_{xu_o} u^o_1 - H^o_{x\varepsilon_o} \\
0 &= g^o_{x_o} x^o_1 + g^o_{y_o} y^o_1 + g^o_{u_o} u^o_1 + g^o_{\varepsilon_o} - \dot{y}^o_o \\
0 &= H^o_{yx_o} x^o_1 + f^o_{y_o} \lambda^o_1 + H^o_{yy_o} y^o_1 + g^o_{y_o} \mu^o_1 + H^o_{yu_o} u^o_1 + H^o_{y\varepsilon_o} + \dot{\mu}^o_o \\
0 &= H^o_{ux_o} x^o_1 + f^o_{u_o} \lambda^o_1 + H^o_{uy_o} y^1_o + g^o_{u_o} \mu^o_1 + H^o_{uu_o} u^1_o + H^o_{u\varepsilon_o}
\end{aligned}
\tag{3.27}
$$

This is a $2n_s$ dimentional system of inhomogeneous, linear differeltial equations with time-dependent coefficients, subject to $2n_f + n_c$ inhomogeneous algebraic constraint equations; their solution will contain $2n_s$ as yet unknown constants of integration, say $x_1^o(0)$ and $\lambda_1^o(T)$. All quantities in (3.27) which are both sub- and super-scripted "o" are known functions of the known zero order solutions.

The first order initial boundary layer problem comes from (3.11) with j = 1. The $2n_s$ equations for the slow variables are mearly quadratures:

$$\begin{aligned} x_1^{i1}(\tau) &= \int_0^\tau f_o^{i1}(\eta)d\eta + x_{o1} \\ \lambda_1^{i1}(\tau) &= -\int_0^\tau H_{x_o}^{i1}(\eta)d\eta + \lambda_1^{i1}(0) \end{aligned} \tag{3.28}$$

where the $n_s$ constants $\lambda_1^{i1}(0)$ are as yet unknown. The remaining equations have the form

$$\begin{aligned} \frac{dy_1^{i1}}{d\tau} &= g_{y_o}^{i1} y_1^{i1} + g_{x_o}^{i1} x_1^{i1} + g_{u_o}^{i1} u_1^{i1} + g_{\varepsilon_o}^{i1} \\ \frac{d\mu_1^{i1}}{d\tau} &= -H_{yy_o}^{i1} y_1^{i1} - g_{y_o}^{i1} \mu_1^{i1} - H_{yx_o}^{i1} x_1^{i1} \\ &\quad - f_{y_o}^{i1} \lambda_1^{i1} - H_{yu_o}^{i1} u_1^{i1} - H_{y\varepsilon_o}^{i1} - H_{yt_o}^{i1} \tau \end{aligned} \tag{3.29}$$

$$0 = H^{i1}_{uy_o} y^{i1}_1 + g^{i1}_{u_o} \mu^{i1}_1 + H^{i1}_{ux_o} x^{i1}_1$$

$$+ f^{i1}_{u_o} \lambda^{i1}_1 + H^{i1}_{uu_o} u^{i1}_1 + H^{i1}_{u\varepsilon_o} + H^{i1}_{ut_o} \tau \qquad (3.29) \text{ cont.}$$

$$y^{i1}_1(0) = y_{o1}$$

This is a $2n_f$ dimensional system of inhomogeneous, linear differential equations with time-dependent coeeficients, subject to $n_c$ inhomogeneous algebraic constraints and $n_f$ boundary conditions. The $n_f$ constants of integration arising in the solution of (3.29) are to be determined by matching with the outer solution. Similarly, the first order terminal boundary layer problem is obtained from (3.14) as

$$\frac{dy^{i2}_1}{d\sigma} = -g^{i2}_{y_o} y^{i2}_1 - g^{i2}_{x_o} x^{i2}_1 - g^{i2}_{u_o} u^{i2}_1 - g^{i2}_{\varepsilon_o}$$

$$\frac{d\mu^{i2}_1}{d\sigma} = H^{i2}_{yy_o} y^{i2}_1 + g^{i2}_{y_o} \mu^{i2}_1 + H^{i2}_{yx_o} x^{i2}_1$$

$$+ f^{i2}_{y_o} \lambda^{i2}_1 + H^{i2}_{yu_o} u^{i2}_1 + H^{i2}_{y\varepsilon_o} + H^{i2}_{yt_o} \tau \qquad (3.30)$$

$$0 = H^{i2}_{uy_o} y^{i2}_1 + g^{i2}_{u_o} \mu^{i2}_1 + H^{i2}_{ux_o} x^{i2}_1$$

$$+ f^{i2}_{u_o} \lambda^{i2}_1 + H^{i2}_{uu_o} u^{i2}_1 + H^{i2}_{u\varepsilon_o} + H^{i2}_{ut_o} \tau$$

$$\mu^{i2}_1(0) = 0$$

with

$$x_1^{i2}(\sigma) = -\int_0^\sigma f_0^{i2}(\eta)d\eta + x_1^{i2}(0)$$

$$\lambda_1^{i2}(\sigma) = \int_0^\sigma H_{x_0}^{i2}(\eta)d\eta \qquad (3.31)$$

The $n_f$ constants of integration arising in the solution of (3.30) and the $n_s$ constants $x_1^{i2}(0)$ are to be determined by matching.

We will now illustrate the procedure of matching and forming composite solutions to first order by working out the details for the slow state x. (This is usually the variable of most interest in applications since it is usually associated most strongly with the performance of the system.) We will assume that we have not matched to zero order as yet, that is (3.15) and (3.16) will not be used, and we will suppose that suitable constants $\tau^*$ and $\sigma^*$ have been found such that (3.25) is satisfied and the boundary layer motions have reached their equilibrium values for $\tau > \tau^*$ and $\sigma > \sigma^*$. For matching, the following are needed: (i) behavior of $x^0$ near $t = 0$ and $t = T$; (ii) behavior of $x^{i1}$ for large $\tau$, and (iii) behavior of $x^{i2}$ for large $\sigma$. The behavior of $x^0$ near $t = 0$ is obtained by expansion as

$$x^0(\varepsilon,t)\Big|_{t\approx 0} = [x_0^0(t) + x_1^0(t)\varepsilon + \,..\, ]_{t\approx 0}$$

$$= x_0^0(0) + \dot{x}_o^o(0)t + \,..\, + [x_1^0(0) + \,..\,]\varepsilon + \,..$$

$$x^0(\varepsilon,t)\Big|_{t\approx 0} = x_0^0(0) + f_o^o(0)t + x_1^0(0)\varepsilon + \,.. \qquad (3.32)$$

where (3.7) and (3.8) were used. Similarly for $x^o$ near $t = T$:

$$x^o(\varepsilon,t)\Big|_{t\approx T} = x_o^o(T) + f_o^o(T)(t-T) + x_1^o(T)\varepsilon + .. \tag{3.33}$$

From (3.10), (3.18), and (3.28), the behavior of $x^{i1}$ at large $\tau$ is

$$x^{i1}(\varepsilon,\tau)\Big|_{\tau>\tau^*} = [x_o^{i1}(\tau) + x_1^{i1}(\tau)\varepsilon + ..]_{\tau>\tau^*}$$

$$= x_{oo} + [\int_0^{\tau^*} f_o^{i1}(\eta)d\eta + f_o^{i1}(\tau^*)(\tau-\tau^*) + x_{o1}]\varepsilon + .. \tag{3.34}$$

$$x^{i1}(\varepsilon,\tau)\Big|_{\tau>\tau^*} = x_{oo} - I_{x_o}^{*i1}\varepsilon + f_o^{i1}(\tau^*)\tau\varepsilon + x_{o1}\varepsilon + ..$$

where

$$I_{x_o}^{*i1} = f_o^{i1}(\tau^*)\tau^* - \int_0^{\tau^*} f_o^{i1}(\eta)d\eta \tag{3.35}$$

are $n_s$ constants similar to $I^*$ shown in Figure 1.2. Similarly, from (3.13), (3.19), and (3.31),

$$x^{i2}(\varepsilon,\tau)\Big|_{\sigma>\sigma^*} = x_o^{i2}(0) + I_{x_o}^{*i2}\varepsilon$$

$$- f_o^{i2}(\sigma^*)\sigma\varepsilon + x_1^{i2}(0)\varepsilon + .. \tag{3.36}$$

where

$$I_{x_o}^{*i2} = f_o^{i2}(\sigma^*)\sigma^* - \int_0^{\sigma^*} f_o^{i2}(\eta)d\eta \tag{3.37}$$

We now match $x^o$ and $x^{il}$ at $t = 0$ according to (1.21):

$$\lim_{\substack{\varepsilon\to 0\\ t\to 0\\ \varepsilon/t\to 0}} \{x^o(\varepsilon,t) - x^{il}(\varepsilon,\tfrac{t}{\varepsilon})\}$$

$$= \lim \{x_o^o(0) + f_o^o(0)t + x_1^o(0)\varepsilon + \dots$$

$$- x_{oo} + I_{x_o}^{*il}\varepsilon - f_o^{il}(\tau^*)t - x_{o1}\varepsilon - \dots\} = 0 \tag{3.38}$$

Equating coefficients of like powers of $\varepsilon^m t^n$ gives

$$\left.\begin{aligned} &x_o^o(0) - x_{oo} = 0 \\ &f_o^o(0) - f_o^{il}(\tau^*) = 0 \\ &x_1^o(0) + I_{x_o}^{*il} - x_{o1} = 0 \end{aligned}\right\} \tag{3.39}$$

from which

$$\left.\begin{aligned} &x_o^o(0) = x_{oo} \\ &x_1^o(0) = x_{o1} - I_{x_o}^{*il} \\ &CP_{x_1}^{il} = x_{oo} + f_o^o(0)t + x_{o1}\varepsilon - I_{x_o}^{*il}\varepsilon \end{aligned}\right\} \tag{3.40}$$

The first of these gives the initial conditions for the slow state variables x in the zero order outer problem, and agrees with (3.15), as expected. The second provides the initial conditions for x in the first order outer problem; note that these conditions are not simply $x_{o1}$. The second condition in (3.39) is a consequence of boundary layer stability and the smoothness of $f(\cdot)$.

Next match $x^o$ and $x^{i2}$ at $t = T$:

$$\lim_{\substack{\varepsilon\to 0\\ t\to T\\ \varepsilon/(T-t)\to 0}} \{x^o(\varepsilon,t) - x^{i2}(\varepsilon,\tfrac{T-t}{\varepsilon})\}$$

$$= \lim \{x_o^o(T) - f_o^o(T)(T-t) + x_1^o(T)\varepsilon + \ldots \tag{3.41}$$

$$- x^{i2}(0) - I_{x_o}^{*i2}\varepsilon + f_o^{i2}(\sigma^*)(T-t) - x_1^{i2}(0)\varepsilon - \ldots\} = 0$$

Equating the coefficients of like powers of $\varepsilon^m t^n$ gives

$$x_o^o(T) - f_o^o(T)T - x_o^{i2}(0) + f_o^{i2}(\sigma^*)T = 0$$

$$f_o^o(T) - f_o^{i2}(\sigma^*) = 0 \tag{3.42}$$

$$x_1^o(T) - I_{x_o}^{*i2} - x_1^{i2}(0) = 0$$

from which

$$x_o^{i2}(0) = x_o^o(T)$$

$$x_1^{i2}(0) = x_1^o(T) - I_{x_o}^{*i2} \tag{3.43}$$

$$CP_{x_1}^{i2} = x_o^o(T) - f_o^o(T)T + f_o^o(T)t + x_1^o(T)\varepsilon$$

There remains only to form the additive composite, which is as follows:

$$x_1^a(\varepsilon,t) = x_o^o(t) + x_1^o(t)\varepsilon + x_o^{i1}(\tfrac{t}{\varepsilon})$$

$$+ x_1^{i1}(\tfrac{t}{\varepsilon})\varepsilon + x_o^{i2}(\tfrac{T-t}{\varepsilon}) + x_1^{i2}(\tfrac{T-t}{\varepsilon})\varepsilon$$

$$- CP_{x_1}^{i1} - CP_{x_1}^{i2}$$

$$x_1^a(\varepsilon,t) = x_o^o(t) + [x_1^o(t) + \int_0^{t/\varepsilon} f_o^{i1}(\eta)d\eta$$

$$+ I_{x_o}^{*i2} - \int_0^{(T-t)\varepsilon} f_o^{i2}(\eta)d\eta - I_{x_o}^{*i2}]\varepsilon$$

$$- f_o^o(0)t + f_o^o(T)(T-t) \tag{3.44}$$

This will give a good approximation to $x(\varepsilon,t)$ for $\varepsilon$ sufficiently small. It is asymptotically correct to first order in $\varepsilon$, that is the neglected terms are of order two or higher in $\epsilon$, written as

$$x(\varepsilon,t) = x_1^a(\varepsilon,t) + O(\varepsilon^2) \tag{3.45}$$

By definition this means

$$\lim_{\varepsilon\to 0^+} \frac{|x(\varepsilon,t) - x_1^a(\varepsilon,t)|}{\varepsilon^2} < \infty \tag{3.46}$$

To summarize, to construct $x_1^a(\varepsilon,t)$ we need to solve four reduced order problems: (i) the $2n_s$ dimensional zero order outer (reduced) problem to get $x_0^o(t)$, (ii) the $2n_s$ dimensional first order problem to get $x_1^o(t)$, (iii) the initial ZOBLE to get $f_0^{i1}(\tau)$, and (iv) the terminal ZOBLE to get $f_0^{i2}(\sigma)$. As a check, it is easy to show, using (3.44) and the matching relations, that

$$x_1^a(\varepsilon,0) = x_{oo} + x_{o1}\varepsilon$$

$$x_1^a(\varepsilon,t)\Big|_{\substack{\tau>\tau^* \\ \sigma>\sigma^*}} = x_0^o(t) + x_1^o(t)\varepsilon$$

as required. The second of these equations is simply a statement that the composite solution sufficiently far away from the boundaries must be identical to the outer solution.

### 3.2. Linear Example

Consider a forced spring-mass (undamped) system on a semi-infinite time interval with unit spring constant and small mass and subject to specified initial conditions. The state equations are

$$\dot{x} = y \qquad x(\varepsilon,0) = \alpha$$

$$\varepsilon\dot{y} = -x + u \qquad y(\varepsilon,0) = \beta$$

It is desired to select the control u(t) to minimize

$$J = \frac{1}{2}\int_0^\infty (y^2 + u^2)dt$$

From (2.8) and (2.9) the adjoint equations and the Hamiltonian function are, with $\lambda_0 = -1$,

$$\dot{\lambda} = \mu$$

$$\varepsilon\dot{\mu} = y - \lambda$$

$$H = -\frac{1}{2}(y^2 + u^2) + y\lambda + (-x + u)\mu$$

For optimal control, $H_u = 0$ from Theorem 2.1 which gives $u = \mu$. Since the system is controllable and observable, $x(\varepsilon,t) \to 0$ and $y(\varepsilon,t) \to 0$ as $t \to \infty$; therefore the 2PBVP to be solved is

$$\begin{aligned} \dot{x} &= y & &; \quad x(\varepsilon,0) = \alpha \\ \varepsilon\dot{y} &= -x + \mu & &; \quad y(\varepsilon,0) = \beta \\ \dot{\lambda} &= \mu & &; \quad \lim_{t\to\infty} x(\varepsilon,t) = 0 \\ \varepsilon\dot{\mu} &= y - \lambda & &; \quad \lim_{t\to\infty} y(\varepsilon,t) = 0 \end{aligned}$$

The matrix G of Theorem 2.2 is

$$G = \begin{pmatrix} 0 & 1 \\ 1 & 0 \end{pmatrix}$$

Since the eigenvalues of this matrix are $\pm 1$, the criterion (2.17) is satisfied and it is obvious that all the hypotheses of Theorem 2.2 are met. The zeroth order approximation to this problem was obtained in [2] by the MAE method; we now obtain the first order approximation.

The outer system is the full system without the boundary conditoons:

$$\dot{x}^o = y^o$$

$$\varepsilon\dot{y}^o = -x^o + \mu^o$$

$$\dot{\lambda}^o = \mu^o$$

$$\varepsilon\dot{\mu}^o = y^o - \lambda^o$$

To solve these first order, set

$$x^o(\varepsilon,t) = x_o^o(t) + \varepsilon x_1^o(t)$$

$$y^o(\varepsilon,t) = y_o^o(t) + \varepsilon y_1^o(t)$$

$$\lambda^o(\varepsilon,t) = \lambda_o^o(t) + \varepsilon\lambda_1^o(t)$$

$$\mu^o(\varepsilon,t) = \mu_o^o(t) + \varepsilon\mu_1^o(t),$$

substitute into the outer equations, and equate coefficients of like powers of $\varepsilon$. The zero order problem is

$$\dot{x}_o^o = y_o^o$$

$$0 = -x_o^o + \mu_o^o$$

$$\dot{\lambda}_o^o = \mu_o^o$$

$$0 = y_o^o - \lambda_o^o$$

The solution for x is

$$x_o^o = c_1 e^t + c_2 e^{-t}$$

Since we know from Section 3.1 that the zero order outer solution is the same as the solution of the reduced problem, we anticipate the matching by setting $c_1 = 0$ and $c_2 = \alpha$. The zero order outer solution is then

$$x_o^o = \mu_o^o = -y_o^o = -\lambda_o^o = \alpha e^{-t}$$

The first order problem is

$$\dot{x}_1^o = y_1^o$$

$$\dot{y}_o^o = -x_1^o + \mu_1^o$$

$$\dot{\lambda}_1^o = \mu_1^o$$

$$\dot{\mu}_o^o = y_1^o - \lambda_1^o$$

The solution for x is

$$x_1^o = c_3 e^t + c_4 e^{-t} - \alpha t e^{-t}$$

In order to match this at the "boundary at $\infty$", we again anticipate the matching by setting $c_3 = 0$. The solution is then

$$x_1^o = -\lambda_1^o = (c_4 - \alpha t)e^{-t}$$

$$y_1^o = -\mu_1^o = -\alpha e^{-t} - (c_4 - \alpha t)e^{-t}$$

Since this outer solution satisfies all the terminal boundary conditions, only an initial boundary layer is required.

The initial boundary layer system is (set $\tau = t/\varepsilon$)

$$\frac{dx^i}{d\tau} = \varepsilon y^i \qquad ; \quad x^i(\varepsilon,0) = \alpha$$

$$\frac{dy^i}{d\tau} = -x^i + \mu^i \qquad ; \quad y^i(\varepsilon,0) = \beta$$

$$\frac{d\lambda^i}{d\tau} = \varepsilon\mu^i$$

$$\frac{d\mu^i}{d\tau} = y^i - \lambda^i$$

To solve these to first order, set

$$x^i(\varepsilon,\tau) = x_o^i(\tau) + \varepsilon x_1^i(\tau)$$

$$y^i(\varepsilon,\tau) = y_o^i(\tau) + \varepsilon y_1^i(\tau)$$

$$\lambda^i(\varepsilon,\tau) = \lambda_o^i(\tau) + \varepsilon\lambda_1^i(\tau)$$

$$\mu^i(\varepsilon,\tau) = \mu_o^i(\tau) + \varepsilon\mu_1^i(\tau)$$

The leading terms give

$$\frac{dx_o^i}{d\tau} = 0 \qquad ; \quad x_o^i(0) = \alpha$$

$$\frac{dy_o^i}{d\tau} = -x_o^i + \mu_o^i \qquad ; \quad y_o^i(0) = \beta$$

$$\frac{d\lambda_o^i}{d\tau} = 0$$

$$\frac{d\mu_o^i}{d\tau} = y_o^i - \lambda_o^i$$

The solution for $y_o^i$ is

$$y_o^i = c_5 e^{\tau} + c_6 e^{-\tau} + c_7$$

Based on the discussion in the previous section, we know that in order to match y, this solution must be asymptotically stable; therefore set $c_5 = 0$ and $c_7 = \beta - c_6$ to get the solution

$$x_o^i = \alpha$$

$$y_o^i = (\beta - c_6)e^{-\tau} + c_6$$

$$\lambda_o^i = c_6$$

$$\mu_o^i = \alpha - (\beta - c_6)e^{-\tau}$$

The first order problem is

$$\frac{dx_1^i}{d\tau} = y_o^i \qquad ; \quad x_1^i(0) = 0$$

$$\frac{dy_1^i}{d\tau} = -x_1^i + \mu_1^i \qquad ; \quad y_1^i(0) = 0$$

$$\frac{d\lambda_1^i}{d\tau} = \mu_o^i$$

$$\frac{d\mu_1^i}{d\tau} = y_1^i - \lambda_1^i$$

The solution is

$$x_1^i = -(\beta - c_6)e^{-\tau} + c_6\tau + (\beta - c_6)$$

$$y_1^i = -(c_6 + c_8)e^{-\tau} + (c_6 + c_8) + \alpha\tau + (\beta - c_6)\tau e^{-\tau}$$

$$\lambda_1^i = \alpha\tau + (\beta - c_6)e^{-\tau} + c_8$$

$$\mu_1^i = (c_6 + c_8)e^{-\tau} + \alpha - (\beta - c_6)\tau e^{-\tau} + c_6\tau + (\beta - c_6)$$

The next step is to match x and $\lambda$ to get the as yet unknown constants of integration $c_4$, $c_6$ and $c_8$. The behavior of $x^o$ at small t is

$$x^o = \alpha e^{-t} + \varepsilon(c_4 - \alpha t)e^{-t}$$

$$\approx \alpha(1 - t) + \varepsilon(c_4 - \alpha t)(1 - t) + \ldots$$

and the behavior of $x^i$ at large $\tau$ is

$$x^i = \alpha + \varepsilon[-(\beta - c_6)e^{-\tau} + c_6\tau + (\beta - c_6)]$$

$$\approx \alpha + \varepsilon[c_6 \frac{t}{\varepsilon} + (\beta - c_6)] + \ldots$$

where the exponential has been neglected because it will be smaller than any algebraic term. Now match according to (3.38):

$$\lim_{\substack{\varepsilon \to 0 \\ t \to 0 \\ \varepsilon/t \to 0}} \{\alpha - \alpha t + \varepsilon c_4 + \ldots - \alpha - c_6 t - \varepsilon(\beta - c_6)\} = 0$$

This implies that we must have

$$c_6 = -\alpha \quad ; \quad c_4 = \alpha + \beta$$

A similar procedure for $\lambda$ gives

$$\lim_{\substack{\varepsilon \to 0 \\ t \to 0 \\ \varepsilon/t \to 0}} \{-\alpha + \alpha t - \varepsilon(\alpha + \beta) + \ldots + \alpha - \alpha t - \varepsilon c_8 - \ldots\} = 0$$

which implies

$$c_8 = -(\alpha + \beta)$$

All constants have now been determined.

The final step is to form composite solutions for the state variables x and y according to (3.44). From the matching relation for x,

$$CP_{x_1} = \alpha - \alpha t + \varepsilon(\alpha + \beta)$$

Then

$$x_1^a(\varepsilon,t) = \alpha e^{-t} + \varepsilon[(\alpha + \beta - \alpha t)e^{-t} - (\alpha + \beta)e^{-t/\varepsilon}]$$

Matching y gives

$$CP_{y_1} = -\alpha + \alpha t - \varepsilon(2\alpha + \beta)$$

so that

$$y_1^a(\varepsilon,t) = -\alpha e^{-t} + (\alpha + \beta)(1 + t)e^{-t/\varepsilon}$$

$$+ \varepsilon[-\alpha e^{-t} - (\alpha + \beta - \alpha t)e^{-t} + (2\alpha + \beta)e^{-t/\varepsilon}]$$

It is easy to see that these composite solutions satisfy all boundary conditions.

Finally, we note that the initial condition on $x_0^1$ could have been obtained directly from (3.39) and (3.34) as follows:

$$c_4 = x_o^1(0) = -I_{x_o}^{*i} = -\lim_{\tau^*\to\infty} [f_o^i(\tau^*)\tau^* - \int_o^{\tau^*} f_o^i(\eta)d\eta] = -\lim_{\tau^*\to\infty} [y_o^i(\tau^*)\tau^* - \int_o^{\tau^*} y_o^i(\eta)d\eta] = -\lim_{\tau^*\to\infty} [(\beta - c_6)\tau^* e^{-\tau^*} + c_6\tau^* - \int_o^{\tau^*} (\beta - c_6)e^{-\eta}d\eta - \int_o^{\tau^*} c_6 d\eta]$$

$$= -\lim_{\tau^*\to\infty} [(\beta - c_6)(e^{-\tau^*}\tau^* + e^{-\tau^*} - 1)] = \beta - c_6 = \alpha + \beta$$

### 3.3. Nonlinear Example

Here we will consider the same example considered in [19] and obtain the solution to first order by the MAE method developed in the previous section. In [19] the solution was obtained by a different but asymptotically equivalent procedure. The state equations are

$$\dot{x} = f = y + u^2 - tv + \varepsilon(x+u)$$

$$\varepsilon\dot{y} = g = -y - tu + 2v^2 + \varepsilon^2(x+y^2)$$

subject to initial conditions

$$x(\varepsilon,0) = x_o(\varepsilon) = x_{00} + x_{01}\varepsilon + \ldots$$

$$y(\varepsilon,0) = y_o(\varepsilon) = y_{00} + y_{01}\varepsilon + \ldots$$

where x and y are scalars, u and v are scalar control variables, and the fixed time interval is $0 \le t \le 1$. It is desired to minimize a function of the final state variables, namely

$$J = x(\varepsilon,1) + \frac{1}{2}\varepsilon\, y(\varepsilon,1)$$

Converting this to an integral performance function gives

$$J = \int_0^{x(\varepsilon,1)} dx + x(\varepsilon,0) + \frac{1}{2}\varepsilon\left[\int_0^{y(\varepsilon,1)} dy + y(\varepsilon,0)\right]$$

$$= \int_0^1 \dot{x}\, dt + \frac{1}{2}\varepsilon \int_0^1 \dot{y} dt + \text{const.}$$

$$J = \int_0^1 (f + \frac{1}{2} g) dt + \text{const.}$$

so that

$$\phi = f + \frac{1}{2} g$$

Taking $\lambda_o = -1$ (until proven otherwise), we form the function H' and the adjoint equations from (2.8) and (2.9)

$$H'(\lambda',\mu',x,y,u,v,\varepsilon,t) = -f - \frac{1}{2} g + f\lambda' + g\mu'$$

$$\dot{\lambda}' = \phi_x - f_x\lambda' - g_x\mu' = f_x + \frac{1}{2} g_x - f_x\lambda' - g_x\mu'$$

$$\dot{\mu}' = \phi_y - f_y\lambda' - g_y\mu' = f_y + \frac{1}{2} g_y - f_y\lambda' - g_y\mu'$$

From Theorem 2.1, for optimal control,

$$H'_u = - f_u - \frac{1}{2} g_u + f_u\lambda' + g_u\mu' = 0$$

$$H'_v = - f_v - \frac{1}{2} g_v + f_v\lambda' + g_v\mu' = 0$$

$$\lambda'(\varepsilon,1) = 0 \; ; \qquad \mu'(\varepsilon,1) = 0$$

Noting that $\lambda'$ always occures in combination $(\lambda'-1)$ and $\mu'$ in combination $(\mu' - \frac{1}{2})$, it is convienient to make the change of variables

$$\lambda = \lambda' - 1 \; ; \qquad \mu = \mu' - \frac{1}{2} \; .$$

The 2PBVP may then be stated as

$$\dot{x} = f = y + u^2 - tv + \varepsilon(x+u)$$

$$\epsilon\dot{y} = g = -y - tu + 2v^2 + \varepsilon^2(x+y^2)$$

$$\dot{\lambda} = -H_x = -\varepsilon\lambda - \varepsilon^2\mu$$

$$\epsilon\dot{\mu} = -H_y = -\lambda + \mu - 2\varepsilon^2 y\mu$$

$$0 = H_u = 2u\lambda + \varepsilon\lambda - t\mu$$

$$0 = H_v = -t\lambda + 4v\mu$$

$$x(\varepsilon,0) = x_{00} + x_{01}\varepsilon + ..$$

$$y(\varepsilon,0) = y_{00} + y_{01}\varepsilon + ..$$

$$\lambda(\varepsilon,1) = -1 \;; \quad \mu(\varepsilon,1) = -\frac{1}{2}$$

where now

$$H(\lambda,\mu,x,y,u,v,\varepsilon,t) = f\lambda + g\mu$$

This is not exactly in the form of (2.11) due to the nonzero terminal conditions on $\lambda$ and $\mu$; however this difference is only a technical one and doesnot make necessary any substaintial changes in the procedure. This is the same system obtained in [19] except that the adjoint variables used there are the negatives of ours.

We begin by forming the matrix

$$[g_y, g_u, g_v] = [-1 + 2\varepsilon^2 y \quad -t \quad 4v]$$

On the reduced solution, for which $\varepsilon = 0$,

$$[g_y, g_u, g_v]_r = [-1 \quad -t \quad 4v_r]$$

which obviously has maximum rank, or rank = 1, on $0 \le t \le 1$. Thus by Theorem 2.3 the alternative procedure for the reduced problem may be employed, and we now do so, both to illustrate the procedure and to obtain information needed to analyze boundary layer stability. The reduced problem is

$$\dot{x}_r = y_r + u_r^2 - tv_r = f_r \; ; \; x_r(0) = x_o$$

$$0 = -y_r - tu_r + 2v_r^2 = g_r$$

$$J_r = \int_0^1 (f_r + \frac{1}{2} g_r)dt$$

Introduce variational multiplier $\lambda_r'$ and ordinary Lagrange multiplier $\mu_r$ and form the function

$$H_r' = -(f_r + \frac{1}{2} g_r) + f_r\lambda_r' + g_r\mu_r'$$

Rescale the multipliers by $\lambda_r = \lambda_r' - 1$ and $\mu_r = \mu_r' - \frac{1}{2}$ as before to get the adjoint equation

$$\dot{\lambda}_r = -H_{x_r} = -\lambda_r f_{x_r} - \mu_r g_{x_r} = 0$$

where now

$$H_r = \lambda_r f_r + \mu_r g_r$$

Thus

$$\lambda_r(t) = \lambda_r(1) = \lambda_r'(1) - 1 = -1.$$

In this problem, there are three control variables, so that

$$0 = H_{y_r} = \lambda_r f_{y_r} + \mu_r g_{y_r} = \lambda_r - \mu_r$$

$$0 = H_{u_r} = 2u_r \lambda_r - t\mu_r$$

$$0 = H_{v_r} = -t\lambda_r + 4v_r \mu_r$$

Solution of these equations gives

$$\lambda_r = -1; \ \mu_r = -1; \quad u_r = t/2$$

$$v_r = t/4; \quad y_r = -\frac{3t^2}{8}; \quad x_r = -\frac{t^3}{8} + x_{oo}$$

We are now in a position to check the eigenvalue criterion (2.17). First, form the constinuents of the matrix G of (2.15) and (2.16):

$$g_y = [-1 + 2\varepsilon^2 y]$$

$$g_u = [-t \qquad 4v]$$

$$H_{uu} = \begin{bmatrix} 2\lambda & 0 \\ 0 & 4\mu \end{bmatrix}$$

$$H_{uu}^{-1} = \begin{bmatrix} \frac{1}{2\lambda} & 0 \\ 0 & \frac{1}{4\mu} \end{bmatrix}$$

$$H_{uy} = [\,0 \quad 0\,]$$

$$H_{yy} = [\,2\varepsilon^2\mu\,]$$

On the reduced solution;

$$g_{y_r} = [-1]\ ; \qquad g_u = [-t \quad 4v_r]$$

$$H_{uu}^{-1} = \begin{bmatrix} \frac{1}{2\lambda_r} & 0 \\ 0 & \frac{1}{4\mu_r} \end{bmatrix} ; \; H_{uy_r} = [\,0 \quad 0\,]$$

$$H_{yy_r} = [\,0\,]$$

Thus, from (2.16)

$$A_r = [\,-1\,]\ ; \quad B_r = \left[\frac{t^2}{2\lambda_r} + \frac{4v_r^2}{\mu_r}\right] ; \quad C_r = [\,0\,]$$

and

$$G_r = \begin{bmatrix} -1 & \frac{t^2}{2\lambda_r} + \frac{4v_r^2}{\mu_r} \\ 0 & 1 \end{bmatrix}$$

The eigenvalues of this matrix are $\pm 1$ and therefore condition (2.17) is satisfied and we can expect boundary layer stability, provided the boundary conditions are close enough to the reduced solution evaluated at the **boundary points to be within their domains of influence.**

We now procede to solve the problem to first order by MAE. The outer problem is

$$\dot{x}^o = y^o + u^{o2} - tv^o + \varepsilon(x^o + u^o)$$

$$\varepsilon\dot{y}^o = -y^o - tu^o + 2v^{o2} + \varepsilon^2(x^o + y^{o2})$$

$$\dot{\lambda}^o = -\varepsilon\lambda^o - \varepsilon^2\mu^o$$

$$\varepsilon\dot{\mu}^o = -\lambda^o + \mu^o - 2y^o\mu^o\varepsilon^2$$

$$0 = \lambda^o(2u^o + \varepsilon) - \mu^o t$$

$$0 = -\lambda^o t + 4\mu^o v^o$$

Set

$$x^o(\varepsilon,t) = x_o^o(t) + x_1^o(t)\varepsilon$$

$$y^o(\varepsilon,t) = y_o^o(t) + y_1^o(t)\varepsilon$$

$$\vdots$$

$$v^o(\varepsilon,t) = v_o^o(t) + v_1^o(t)\varepsilon$$

and substitute into the outer system and equate terms with like powers of $\varepsilon$. The zero order outer problem is

$$\dot{x}_o^o = y_o^o + u_o^{o2} - tv_o^o$$

$$0 = -y_o^o - tu_o^o + 2v_o^{o2}$$

$$\dot{\lambda}_o^o = 0$$

$$0 = -\lambda_o^o + \mu_o^o$$

$$0 = 2\lambda_o^o u_o^o - \mu_o^o t = 0$$

$$0 = -\lambda_o^o t + 4\mu_o^o v_o^o = 0$$

with solution (assuming we don't know the boundary conditions)

$$\lambda_o^o = c_1; \quad \mu_o^o = c_1; \quad u_o^o = \frac{t}{2}; \quad v_o^o = \frac{t}{4}$$

$$y_o^o = -\frac{3}{8}t^2; \quad x_o^o = -\frac{1}{8}t^3 + c_2$$

where $c_1$ and $c_2$ are constants of integration. The first order problem is

$$\dot{x}_1^o = y_1^o + 2u_o^o u_1^o - tv_1^o + x_o^o + u_o^o$$

$$\dot{y}_o^o = -y_1^o - tu_1^o + 4v_o^o v_1^o$$

$$\dot{\lambda}_1^o = -\lambda_o^o$$

$$\dot{\mu}_o^o = -\lambda_1^o + \mu_1^o$$

$$0 = 2(\lambda_1^o u_o^o + \lambda_o^o u_1^o) + \lambda_o^o - t\mu_1^o$$

$$0 = -t\lambda_1^o + 4(\mu_o^o v_1^o + \mu_1^o v_o^o) = 0$$

As expected, this is a second order linear system in $x_1^o$ and $\lambda_1^o$; its solution is

$$\lambda_1^o = -c_1 t + c_3; \quad \mu_1^o = -c_1 t + c_3$$

$$u_1^o = -\frac{1}{2}; \quad v_1^o = 0$$

$$y_1^o = \frac{5}{4} t; \quad x_1^o = \frac{5}{8} t^2 - \frac{1}{32} t^4 + c_2 t + c_4$$

The initial boundary layer equation is obtained by introducing the stretching transformation (3.3):

$$\frac{dx^{i1}}{d\tau} = \varepsilon[y^{i1} + u^{i2} - \tau\varepsilon v^{i1} + \varepsilon(x^{i1} + u^{i1})]$$

$$\frac{dy^{i1}}{d\tau} = -y^{i1} - \tau\varepsilon u^{i1} + 2v^{i1^2} + \varepsilon^2(x^{i1} + y^{i1^2})$$

$$\frac{d\lambda^{i1}}{d\tau} = -\varepsilon^2\lambda^{i1} - \varepsilon^3\mu^{i1}$$

$$\frac{d\mu^{i1}}{d\tau} = -\lambda^{i1} + \mu^{i1} - 2\varepsilon^2 y^{i1}\mu^{i1}$$

$$0 = 2u^{i1}\lambda^{i1} + \varepsilon\lambda^{i1} - \tau\varepsilon\mu^{i1}$$

$$0 = -\tau\varepsilon\lambda^{i1} + 4v^{i1}\mu^{i1}$$

$$x^{i1}(\varepsilon,0) = x_o(\varepsilon) = x_{oo} + x_{o1}\varepsilon + ..$$

$$y^{i1}(\varepsilon,0) = y_o(\varepsilon) = y_{oo} + y_{o1}\varepsilon + ..$$

To solve there, set

$$x^{il}(\varepsilon,\tau) = x_o^{il}(\tau) + x^{il}(\tau)\varepsilon$$

$$\vdots$$

$$v^{il}(\varepsilon,\tau) = v_o^{il}(\tau) + v_1^{il}(\tau)\varepsilon$$

and equate terms of like powers in $\varepsilon$. The zero order problem is the ZOBLE:

$$\frac{dx_o^{il}}{d\tau} = 0; \quad \frac{dy_o^{il}}{d\tau} = -y_o^{il} + 2{v_o^{il}}^2$$

$$\frac{d\lambda_o^{il}}{d\tau} = 0; \quad \frac{d\mu_o^{il}}{d\tau} = -\lambda_o^{il} + \mu_o^{il}$$

$$0 = 2u_o^{il}\lambda_o^{il}; \quad 0 = 4v_o^{il}\mu_o^{il}$$

$$x_o^{il}(0) = x_{oo}; \qquad y_o^{il}(0) = y_{oo}$$

This is a fourth order system and therefore its solution will have four constants of integration. Two of the equations reduce to quadratures and are thus trivial to solve. Of the remaining two "modes", one will be unstable. The procedure is to use two of the constants of integration to satisfy the two boundary conditions and a third to suppress the instability. This leaves one free constant to be determined by matching with the outer solution. The solution to the ZOBLE which satisfies the two boundary conditions is

$$x_o^{il} = x_{oo} \; ; \; \lambda_o^{il} = c_5$$

$$u_o^{il} = 0 \quad ; \; v_o^{il} = 0$$

$$\mu_o^{il} = c_5 + c_6 e^{\tau} \; ; \; y_o^{il} = y_{oo} e^{-\tau}$$

But for asymptotic stability we must have $c_6 = 0$ and thus $\mu_o^{il} = c_5$. The first order problem is

$$\frac{dx_1^{il}}{d\tau} = y_o^{il} + u_o^{il^2} = y_{oo} e^{-\tau}; \quad x_1^{il}(0) = x_{o1}$$

$$\frac{dy_1^{il}}{d\tau} = -y_1^{il} - \tau u_o^{il} + 4 v_o^{il} v_1^{il} = -y_1^{il} \; ; \quad y_1^{il}(0) = y_{o1}$$

$$\frac{d\lambda_1^{il}}{d\tau} = 0$$

$$\frac{d\mu_1^{il}}{d\tau} = -\lambda_1^{il} + \mu_1^{il}$$

$$0 = 2(u_o^{il}\lambda_1^{il} + u_1^{il}\lambda_o^{il}) + \lambda_o^{il} - \tau\mu_o^{il} = c_5(2u_1^{il} + 1 - \tau)$$

$$0 = -\tau\lambda_o^{il} + 4(v_o^{il}\mu_1^{il} + v_1^{il}\mu_o^{il}) = c_5(-\tau + 4v_1^{il})$$

with solution

$$v_1^{il} = \frac{\tau}{4} \quad ; \qquad u_1^{il} = \frac{\tau - 1}{2}$$

$$\lambda_1^{il} = c_8 \; ; \quad x_1^{il} = y_{oo}(1 - e^{-\tau}) + x_{o1}$$

$$y_1^{i1} = y_{o1}e^{-\tau} \ ; \quad \mu_1^{i1} = c_7 e^{\tau} + c_8$$

This contains two types of terms which grow with $\tau$. The algebraic terms may be needed to match with like terms in the outer solution and we retain them. The exponentially growing term, however, cannot be matched by any algebraic term and must therefore be suppressed by setting $c_7 = 0$, giving $\mu_1^{i1} = c_8$.

The terminal boundary layer equation is obtained by using the transformation (3.5):

$$\frac{dx^{i2}}{d\sigma} = -\varepsilon[y^{i2} + u^{i2^2} - (1 - \varepsilon\sigma)v^{i2} + \varepsilon(x^{i2} + u^{i2})]$$

$$\frac{dy^{i2}}{d\sigma} = -[-y^{i1} - (1 - \varepsilon\sigma)u^{i2} + 2v^{i2^2} + \varepsilon^2(x^{i2} + y^{i2^2})]$$

$$\frac{d\lambda^{i2}}{d\sigma} = -\varepsilon[-\varepsilon\lambda^{i2} - \varepsilon^2\mu^{i2}]$$

$$\frac{d\mu^{i2}}{d\sigma} = -[-\lambda^{i2} + \mu^{i2} - 2\varepsilon^2 y^{i2}\mu^{i2}]$$

$$0 = 2u^{i1}\lambda^{i2} + \varepsilon\lambda^{i2} - (1 - \varepsilon\sigma)\mu^{i2}$$

$$0 = -(1 - \varepsilon\sigma)\lambda^{i2} + 4v^{i2}\mu^{i2}$$

$$\lambda^{i2}(\varepsilon,0) = -1 \ ; \qquad \mu^{i2}(\varepsilon,0) = -\frac{1}{2}$$

To solve to first order, set

$$x^{i2}(\varepsilon,\sigma) = x_o^{i2}(\sigma) + x_1^{i2}(\sigma)\varepsilon$$

$$\vdots$$

$$v^{i2}(\varepsilon,\sigma) = v_o^{i2}(\sigma) + v_1^{i2}(\sigma)\varepsilon$$

The ZOBLE is

$$\frac{dx_o^{i2}}{d\sigma} = 0 \; ; \quad \frac{dy_o^{i2}}{d\sigma} = y_o^{i2} + u_o^{i2} - 2v_o^{i2^2}$$

$$\frac{d\lambda_o^{i2}}{d\sigma} = 0 \; ; \quad \frac{d\mu_o^{i2}}{d\sigma} = \lambda_o^{i2} - \mu_o^{i2}$$

$$0 = 2u_o^{i2}\lambda_o^{i2} - \mu_o^{i2} \; ; \quad 0 = -\lambda_o^{i2} + 4v_o^{i2}\mu_o^{i2}$$

$$\lambda_o^{i2}(0) = -1 \; ; \quad \mu_o^{i2}(0) = -\frac{1}{2}$$

The solution which satisfies the boundary conditions is

$$\lambda_o^{i2} = -1 \quad ; \quad x_o^{i2} = c_9$$

$$\mu_o^{i2} = -1 + \frac{1}{2}e^{-\sigma} \; ; \quad v_o^{i2} = -\frac{1}{4(-1+\frac{1}{2}e^{-\sigma})}$$

$$u_o^{i2} = \frac{1}{2} - \frac{1}{4}e^{-\sigma}$$

$$y_o^{i2} = c_{10}e^{\sigma} - \frac{1}{2} + \frac{1}{8}e^{-\sigma} + \frac{e^{\sigma}}{4(1 - \frac{1}{2}e^{-\sigma})}$$

To make this solution stable we must pick $c_{10} = -\frac{1}{4}$; this gives

$$y_o^{i2} = -\frac{1}{2} + \frac{1}{8}e^{-\sigma} + \frac{1}{8 - 4e^{-\sigma}}$$

The first order problem is

$$\frac{dx_1^{i2}}{d\sigma} = -y_o^{i2} + u_o^{i2^2} + v_o^{i2}$$

$$\frac{dy_1^{i2}}{d\sigma} = y_1^{i2} + u_1^{i2} - \sigma u_o^{i2} - 4v_o^{i2}v_1^{i2}$$

$$\frac{d\lambda_1^{i2}}{d\sigma} = 0$$

$$\frac{d\mu_1^{i2}}{d\sigma} = \lambda_1^{i2} - \mu_1^{i2}$$

$$0 = 2u_o^{i2}\lambda_1^{i2} + 2u_1^{i2}\lambda_o^{i2} + \lambda_o^{i2} - \mu_1^{i2} + \sigma\mu_o^{i2}$$

$$0 = -\lambda_1^{i2} + \sigma\lambda_o^{i2} + 4v_o^{i2}\mu_1^{i2} + 4v_1^{i2}\mu_o^{i2}$$

$$\lambda_1^{i2}(0) = 0; \quad \mu_1^{i2}(0) = 0$$

with solution

$$\lambda_1^{i2} = 0 \; ; \quad \mu_1^{i2} = 0$$

$$u_1^{i2} = \frac{-1 - \sigma + \frac{1}{2}\sigma e^{-\sigma}}{2} \; ; \quad v_1^{i2} = \frac{\sigma}{2(e^{-\sigma} - 2)}$$

$$x_1^{i2} = \frac{3}{8}\sigma - \frac{1}{8}e^{-\sigma} + \frac{1}{32}e^{-2\sigma} + \frac{1}{8}\ln(1 - \frac{1}{2}e^{-\sigma}) + c_{11}$$

$$y_1^{i2} = c_{12}e^{\sigma} + \frac{3}{2} + \sigma - \frac{(2\sigma + 1)e^{-\sigma}}{8} + \frac{1}{2}e^{\sigma}\ln(2 - e^{-\sigma})$$

$$+ \frac{\sigma}{2(e^{-\sigma} - 2)}$$

As before, anticipating the matching requirement, we select $c_{12}$ to cancel the exponentially growing terms; thus set $c_{12} = -\frac{1}{2}\ln 2$ to get

$$y_1^{i2} = \frac{1}{2}e^{\sigma}\ln\frac{2 - e^{-\sigma}}{2} + \frac{3}{2} + \sigma$$

$$- \frac{(2\sigma + 1)e^{-\sigma}}{8} + \frac{\sigma}{2(e^{-\sigma} - 2)}$$

The next step is to match the outer and boundary layer solutions to get the unknown constants of integration and the common parts needed for construction of the composite solutions. Only the slow variables, in this case x and $\lambda$, need be matched to get the constants.

Starting with the matching of x, we know that we will need the limiting behavior of the outer and inner expansions of x as given by (3.32), (3.33), (3.34) and (3.36). The required expressions are, letting t' = t -

$$x_o^o(t)\Big|_{t\approx o} \approx c_2 \qquad x_o^o(t)\Big|_{t\approx 1} \approx -\frac{1}{8}(1-3t') + c_2$$

$$x_1^o(t)\Big|_{t\approx o} \approx c_4 \qquad x_1^o(t)\Big|_{t\approx 1} = \frac{19}{32} + c_2 + c_4$$

$$x_o^{i1}(\tau)\Big|_{\tau>\tau^*} \approx x_{oo} \qquad x_1^{i1}(\tau)\Big|_{\tau>\tau^*} \approx y_{oo} + x_{o1}$$

$$x_o^{i2}(\sigma)\Big|_{\sigma>\sigma^*} \approx c_9 \qquad x_1^{i2}(\sigma)\Big|_{\sigma>\sigma^*} \approx \frac{3}{8}\sigma + c_{11}$$

Matching at t = 0 according to (3.38),

$$\lim_{\substack{\varepsilon\to 0\\ t\to 0\\ \varepsilon/t\to 0}} \{c_2 + c_4\varepsilon - x_{oo} - (y_{oo} + x_{o1})\varepsilon\} = 0$$

which gives

$$c_2 = c_{oo}\ ; \quad c_4 = y_{oo} + x_{o1}; \quad CP_{x_1}^{i1} = x_{oo} + (y_{oo} + x_{o1})\varepsilon$$

Similarly, from (3.41),

$$\lim_{\substack{\varepsilon\to 0\\ t'\to 0\\ \varepsilon/t'\to 0}} \{-\frac{1}{8}(1 - 3t') + c_2 + \varepsilon(\frac{19}{32} + c_2 + c_4) - c_9 - \varepsilon(\frac{3}{8}\frac{t'}{\varepsilon} + c_{11})\} = 0$$

which gives

$$c_9 = x_{oo} - \frac{1}{8} ; \quad c_{11} = \frac{19}{32} + x_{oo} + y_{oo} + x_{o1}$$

$$CP^{i2}_{x_1} = - \frac{1}{8} + \frac{3}{8} t' + x_{oo} + \varepsilon(\frac{19}{32} + x_{oo} + y_{oo} + x_{o1})$$

Finally, from (3.44)

$$x_1^a(\varepsilon,t) = x_{oo} - \frac{1}{8} t^3 + [x_{o1} + y_{oo} + x_{oo} t$$

$$+ \frac{5}{8} t^2 - \frac{1}{32} t^4 - y_{oo} e^{-\tau} + \frac{1}{8} \ln (1 - \frac{1}{2} e^{-\sigma})$$

$$- \frac{1}{8} e^{-\sigma} + \frac{1}{32} e^{-2\sigma} ]\varepsilon$$

This is the same expression derived in [19] (note the many missprints in the equations on page 324 of [19]).

Turning now to $\lambda$, we have

$$\lambda_o^o(t)\Big|_{t\approx o} \approx c_1 \qquad \lambda_o^o(t)\Big|_{t\approx 1} \approx c_1$$

$$\lambda_1^o(t)\Big|_{t\approx o} \approx c_3 \qquad \lambda_1^o(t)\Big|_{t\approx 1} \approx -c_1 + c_3$$

$$\lambda_1^o(\tau)\Big|_{\tau>\tau^*} \approx c_5 \qquad \lambda_1^{i1}(\tau)\Big|_{\tau>\tau^*} \approx c_8$$

$$\lambda_o^{i2}(\sigma)\Big|_{\sigma>\sigma^*} \approx -1 \qquad \lambda_1^{i2}(\sigma)\Big|_{\sigma>\sigma^*} \approx 0$$

Matching at $t = 0$ gives

$$c_1 = c_5; \quad c_3 = c_8; \quad CP^{i1}_{\lambda_1} = c_1 + \varepsilon c_3$$

Matching at t = 1 gives

$$c_1 = -1; \quad c_3 = -1; \quad CP^{i2}_{\lambda_1} = -1$$

Forming the first order composite,

$$\lambda^a_1(\varepsilon,t) = -1 + \varepsilon(t-1)$$

which agrees with [19]. All the constants of integration arising in the outer solution now have been determined.

There remains to consider the variables $y,\mu,u$, and $v$. The only reason for matching them is to get the common parts. The result is

$$CP^{i1}_{y_1} = 0; \quad CP^{i2}_{y_1} = \frac{3}{8} - \frac{3}{4}\,t + \frac{5}{4}\,\varepsilon$$

$$CP^{i1}_{\mu_1} = -1 - \varepsilon\;; \quad CP^{i2}_{\mu_1} = -1$$

$$CP^{i1}_{u_1} = \frac{1}{2}(t-\varepsilon); \quad CP^{i2}_{u_1} = \frac{1}{2}\,(t-\varepsilon)$$

$$CP^{i1}_{v_1} = \frac{t}{4}\;; \quad CP^{i2}_{v_1} = \frac{t}{4}$$

The composite solutions are then

$$y_1^a(\varepsilon,t) = y_{oo}e^{-\tau} - \frac{3}{8}t^2 - \frac{1}{8} + \frac{1}{8}e^{-\sigma} + \frac{1}{8-4e^{-\sigma}}$$

$$+ \varepsilon[\frac{5}{4}t + \frac{1}{2}e^{\sigma}\ln(1 - \frac{1}{2}e^{-\sigma}) - \frac{(2\sigma+1)}{8}e^{-\sigma}$$

$$+ \frac{\sigma}{2(e^{-\sigma}-2)} + \frac{\sigma}{4} + \frac{1}{4}]$$

$$\mu_1^a(\varepsilon,t) = -1 + \frac{1}{2}e^{-\sigma} + \varepsilon(t-1)$$

$$u_1^a(\varepsilon,t) = \frac{t}{2} - \frac{t}{4}e^{-\sigma} - \frac{1}{2}\varepsilon$$

$$v_1^a(\varepsilon,t) = \frac{t}{2(2-e^{-\sigma})}$$

which also agree with [19] (allowing for missprints), with the possible exception of $y_1^a(\varepsilon,t)$.

ACKNOWLEDGMENT

This paper was written while the author was a Visiting Research Fellow at the Twente University of Technology, The Netherlands. The support of the University is gratefully acknowledged.

REFERENCES

[1] Wasow, W.R., Asymptotic expansions for ordinary differential equations, *Interscience*, 1965.

[2] Ardema, M.D., Singular perturbations in flight mechanics, *NASA TM X-62*, 380, July 1977.

[3] Ardema, M.D., Solution of second order linear system by matched asymptotic expansions, NASA TM 84, 246, to appear.

[4] Tihonov, A.N., Systems of differential equations containing small parameters multiplying some of the derivatives, *Math. Sb.*, vol. 73, no. 3, N.S. (31), 1952 (in Russian).

[5] Vasileva, A.B., Asymptotic behavior of solutions to certain problems involving nonlinear differential equations containing a small parameter multiplying the highest derivatives, *Russian Math. Surveys*, vol. 18, no. 3, 1963 (English translation).

[6] Levin, J.J., and Levinson, N., Singular perturbations of non-linear systems of differential equations and an associated boundary layer equation, *J. Rat. Mech. Anal.*, vol. 3, 1954.

[7] O'Malley, R.E., Introduction to singular perturbations, *Academic*, 1974.

[8] Buti G., The singular perturbation theory of differential equations in control theory, *Periodica Polytechnica*, vol. 10, no. 2, 1965.

[9] Cole, J.D., Perturbations methods in applied mathematics, *Blaisdell*, 1968.

[10] Van Dyke, M., Perturbations methods in fluid mechanics, *Academic*, 1964.

[11] Nayfeh, A.H., Perturbations methods, *Wiley*, 1973.

[12] Eckhaus, W., Matched asymptotic expansions and singular perturbations, *North-Holland*, 1973.

[13] Pontryagin, L.S., Boltyanskii, V.G., Gamkrelidze, R.V., and Mishchenko, E.F., The mathematical theory of optimal processes, *Interscience*, 1962.

[14] Leitmann, G., An introduction to optimal control, *McGraw-Hill*, 1969.

[15] Bryson, A.E., jr. and Ho, Y.-C., Applied optimal Control, *Blaisdell*, 1968.

[16] Hadlock, C.R., Singular perturbations of a class of two point boundary value problems arising in optimal control, Coordinated Science Laboratory, *Report R-481*, Univ. of Illinois, July 1970.

[17] Hadlock, C.R., Existence and dependence on a parameter of solutions of a nonlinear two point boundary value problem, *Journal Differential Equations*, vol. 14, no. 3, November 1973.

[18] Freedman, M.I., and Kaplan, J.L., Singular perturbations of two-point boundary value problems arising in optimal control, *SIAM Journal Control and Optimization*, vol. 14, no. 2, February 1976.

[19] Freedman, M.I., and Granoff, B., Formal asymptotic solution of a singularly perturbed nonlinear optimal control problem, *Journal Optimization Theory and Applications*, vol. 19, no. 2, June 1976.

[20] O'Malley, R.E., Singular perturbations and optimal control, Lectures given at the conference on Mathematical Control Theory, Australian National University, Canberrra, August 23 - September 2, 1977.

[21] Ardema, M.D., Characteristics of the boundary layer equations of the minimum time-to-climb problem, Proceedings of the Fourteeth Annual Allerton Conference on Circuit and System Theory, September 1976.

[22] Sacher, R.J., and Sell, G.R., Singular perturbations and conditional stability, *J. of Math. Analysis and Applic.*, vol. 76, 1980.

[23] Berger, S.A., Laminar wakes, *American Elsevier*, 1971.

[24] Ai, D.K., On the critical point of the Crocco-Lees mixing theory in the laminar near wake, *J. Eng. Math.*, vol. 4, no. 2, April 1970.

[25] O'Malley, R.E., Introduction to singular perturbations, *Academic*, 1974.

[26] Kokotovic, P.V., O'Malley, R.E., and Sannuti, P., Singular perturbations and order reduction in control theory - an overview, *Automatica*, vol. 12, 1976.

[27] Kokotovic, P.V. (ed.), Singular perturbations and time scales in modeling and control of dynamic systems, Coordinated Science Laboratory, *Report R-901*, Univ. of Illinois, November 1980.

[28] O'Malley, R.E., On the asymptotic solution of certain non-linear singularly perturbed optimal control problems, Order Reduction in Control System Design, *ASME*, 1972.

[29] Chow, J.H., A class of singularly perturbed, nonlinear fixed-end-point control problems, *JOTA*, vol. 29, no. 2, October 1979.

[30] Ardema, M.D., Solution of the minimum time-to-climb problem by matched asymptotic expansions, *AIAA Journal*, vol. 14, no. 7, July 1976.

[31] Ardema, M.D., Linearization of the boundary-layer equations of the minimum time-to-climb problem, *Journal Guidance and Control*, vol. 2, no. 5, September - October, 1979.

[32] Kelley, H.J., Singular perturbations for a Mayer variational problem, *AIAA Journal*, vol. 8, no. 6, June 1970.

[33] Kao, Y.K., and Bankoff, S.G., Singular perturbations analysis and free-time optimal control problems, Paper 7-3 presented at the Joint Automatic Control Conference, Columbus, Ohio, June 20-22, 1973.

[34] Calise, A.J., and Aggarwal, R., A Conceptual approach to applying singular perturbation methods to variational problems, Proceedings of the Eleventh Annual Allerton Conference on Circuit and Systems Theory, Allerton, 1973.

[35] Calise, A.J., Extended energy management methods for flight performance optimization, *AIAA Journal*, vol. 15, no. 3, March 1977.

[36] Visser, H.G., and Shinar, J., Asymptotic solution of a problem of missile guidance by proportional navigation for a singularly perturbed system, *Technion Report TAE no. 464*, Haifa, Israël, October, 1981.

# ON NONLINEAR OPTIMAL CONTROL PROBLEMS

Robert E. O'Malley, Jr.
Department of Mathematical Sciences
Rensselaer Polytechnic Institute
Troy, New York 12181
U.S.A.

Some of the first papers which consciously used a singular perturbations approach in optimal control were about nonlinear regulators (cf. Kokotovic and Sannuti (1968) and Sannuti and Kokotovic (1969)). They showed how such problems can be reduced to nonlinear two-point singularly perturbed boundary value problems for the states and costates, constrained by an optimality condition. Even without such constraints, however, a general theory for such nonlinear systems of ordinary differential equations is not available (cf. O'Malley (1980)). Limited success has been achieved, but largely for quasilinear problems (cf., e.g., Hadlock (1973), O'Malley (1974), Sannuti (1975), Freedman and Granoff (1976), and Freedman and Kaplan (1976) and (1977)). Efficient numerical methods are now becoming available for such boundary value problems (cf., e.g., Flaherty and O'Malley

(1982), Weiss (1982), and Ascher and Weiss (1982)). For more nonlinear problems, interior shocks and transition layers can be expected in addition to endpoint boundary layers (cf. Howes (1978) for a treatment of second order scalar equations and O'Malley (1982) concerning some special vector systems). One must anticipate that such theories will become important in various control contexts. In addition, the associated mathematics may become increasingly geometrical (cf. Levin and Levinson (1954), Fenichel (1979), Sastry, Desoer, and Varaiya (1980), and Kurland (1981)). One would hope that the underlying Hamiltonian structure of control problems might ultimately allow a special development of the necessary asymptotic analysis. Our approach will largely follow McIntyre (1977) (see O'Malley (1978)).

Suppose that we wish to minimize a scalar cost

$$J = \lambda(x(1),z(1),\varepsilon) + \int_0^1 \Lambda\,(x(t),z(t),u(t),t,\varepsilon)\,dt \tag{1}$$

subject to the state constraints

$$\dot{x} = f(x,z,u,t,\varepsilon) \tag{2}$$

$$\varepsilon\dot{z} = g(x,z,u,t,\varepsilon) \tag{3}$$

where x,z, and u are vectors of dimensions m,n, and r, respectively, $\varepsilon$ is a small positive parameter, and x(0) and z(0) are prescribed. We shall assume smoothness of all coefficients, and seek the limiting solution as $\varepsilon \to 0$. Fast dynamics can be expected whenever $\dot{z} = O(\frac{1}{\varepsilon})$. We note, curiously, that the corresponding infinite interval problems might be more tractable, because the solutions would be expressed in terms of singularly perturbed initial value problems. Their nonlinear theory is much more developed than that for two-point problems (cf. Hoppensteadt (1971) and

O'Malley (1974)).

Necessary conditions for optimality follow from, for example, the Pontryagin maximum principle. Thus, we attempt to minimize the hamiltonian

$$h = \Lambda + p'f + q'g \tag{4}$$

where the costates p and $\varepsilon q$ satisfy

$$\dot{p} = -h_x, \quad p(1,\varepsilon) = \lambda_x(x(1), z(1), \varepsilon) \tag{5}$$

$$\varepsilon\dot{q} = -h_z, \quad \varepsilon q(1,\varepsilon) = \lambda_z(x(1), z(1), \varepsilon) \tag{6}$$

(corresponding to the state equations (2) and (3) which can now be rewritten as $\dot{x} = h_p$ and $\varepsilon\dot{z} = h_q$). Optimality requires that

$$h_u = 0 \tag{7}$$

and the Legendre-Clebsch condition

$$h_{uu} \geq 0 \tag{8}$$

should hold at the minimum.

Introducing

$$\psi = \begin{pmatrix} x \\ p \end{pmatrix} \text{ and } \zeta = \begin{pmatrix} z \\ q \end{pmatrix} \tag{9}$$

and identifying $H(\psi,\zeta,u,t,\varepsilon)$ with $h(x,z,u,p,q,t,\varepsilon)$, conditions (5)-(7) take the form

$$\left.\begin{aligned} \dot{\psi} &= J_m H_\psi \\ \varepsilon\dot{\zeta} &= J_n H_\zeta \\ H_u &= 0 \end{aligned}\right\} \tag{10}$$

where $J_k$ is the symplectic matrix

$$J_k = \begin{pmatrix} 0 & I_k \\ -I_k & 0 \end{pmatrix}$$

of dimension $2k \times 2k$. When the strong form of the Legendre-Clebsch condition (8) holds and the matrix $H_{uu}$ is positive definite, the implicit function theorem guarantees that the optimality condition $H_u = 0$ can be uniquely solved (locally) in the form

$$u = \eta(\psi,\zeta,t,\varepsilon), \tag{11}$$

leaving a singularly perturbed system

$$\left.\begin{aligned} \dot{\psi} &= F(\psi,\zeta,t,\varepsilon) \equiv J_m \, H_\psi(\psi,\zeta,\eta,t,\varepsilon) \\ \varepsilon\dot{\zeta} &= G(\psi,\zeta,t,\varepsilon) \equiv J_n \, H_\zeta(\psi,\zeta,\eta,t,\varepsilon) \end{aligned}\right\} \tag{12}$$

of dimension $2m + 2n$ subject to the $m + n$ initial conditions specified by $x(0)$ and $z(0)$ and the $m + n$ terminal conditions for $p$ and $q$ given by (5) and (6). The system (12) depends, of course, on which root (11) of $H_u = 0$ is selected, whenever $h$ is superquadratic in $u$. More attention should be given to problems where non-smooth controls result from switching from one root of (11) to another at points within our interval $[0,1]$.

Solutions to singularly perturbed systems of the form (12) depend critically on the location of the eigenvalues of $G_\zeta$. Since $H_u = 0$ determines $u$ and $\frac{du}{d\zeta}$, we obtain

$$G_\zeta = J_n(H_{\zeta\zeta} - H'_{u\zeta}H_{uu}^{-1}H_{u\zeta}) \tag{13}$$

Since $G_\zeta$ is Hamiltonian (with $J_nG_\zeta = -G'_\zeta J_n$ symmetric), eigenvalues of $G_\zeta$ occur in pairs $\pm\lambda$. For this reason, we shall call our problem conditionally stable, noting that rapidly growing modes of the linearized problem are paired with rapidly decaying modes. Away from boundary and interior regions of nonuniform convergence, solutions to the system (12) will be well-approximated by solutions of the limiting system

$$\left.\begin{aligned} \dot{\psi} &= F(\psi,\zeta,t,0) \\ 0 &= G(\psi,\zeta,t,0) \end{aligned}\right\} \tag{14}$$

If we now assume that $G_\zeta$ is nonsingular, we will be able to solve the second system in a locally unique manner as

$$\zeta = \xi(\psi,t) \tag{15}$$

so the reduced problem is described through a $2n^{th}$ order system

$$\dot{\psi} = \mathcal{F}(\psi,t) \equiv F(\psi,\xi(\psi,t),t,0). \tag{16}$$

Different roots $\xi$ of $G = 0$ will, of course, result in different limiting systems.

The simplest possibility is to assume that $G_\zeta$ always has n eigenvalues $\lambda$ strictly in the left half-plane (with the remaining n eigenvalues $-\lambda$ necessarily being in the right half plane). This strong assumption eliminates turning points, and assures us that transition layers will have $O(\varepsilon)$ thickness, corresponding to local stretched variables of the form $\kappa = (x-\tilde{x})/\varepsilon$. Assuming further that only endpoint boundary layers develop, we can expect to "lose" m of the m + n boundary conditions at t = 0 and at t = 1 in the limit as $\varepsilon \to 0$. Simple cancellation laws as for linear problems (cf., e.g., Harris (1973)) cannot generally be expected, however. Moreover, the asymptotically unbounded terminal vector for $q(1,\varepsilon)$ (cf. (6)) suggests the possibility of $O(\frac{1}{\varepsilon})$ impulses occuring in the solution at t = 1 (cf. O'Malley (1980) and Glizer and Dmitriev (1975, 1979)). If we, however, assume that the terminal cost function $\lambda(x,z,\varepsilon)$ in (1) is a slowly-varying function $\phi(x,\varepsilon z,\varepsilon)$, we can expect a bounded solution throughout [0,1] featuring nonuniform convergence in the fast, $\zeta$, variable at both endpoints. (cf. O'Malley (1970)). Indeed, the limiting solution will then be found in terms of the reduced two-point boundary value problem

$$\left.\begin{array}{l} \dot{\psi} = F(\psi,t) \\ \text{with } x(0) \text{ prescribed and} \\ p(1) = \theta_x(x(1),0,0) \end{array}\right\} \tag{17}$$

Corresponding to any such solution $\psi$, the nonuniform convergence at $t = 0$ would be obtained through an initial layer coordinate $\tau = \frac{t}{\varepsilon}$ and a conditionally stable initial layer system

$$\left.\begin{array}{l} \frac{dv}{d\tau} = G(\psi(0),\xi(\psi(0),0) + v,0,0), \ \tau \geq 0 \\ v(0) = \zeta(0) - \xi(\psi(0),0). \end{array}\right\} \tag{18}$$

An analogous terminal layer occurs at $t = 1$. Moreover, asymptotic expansions of optimal states, control, and cost can be obtained in a straightforward manner.

It is generally very difficult to provide verifiable hypotheses which guarantee the decay of solutions to such layer systems for an appropriate domain of initial vectors. Generally, however, the forms of possible asymptotic solutions will be much richer than those traditionally used for such linear systems (without turning points). It is essential for progress that we now make the effort to generate numerical and analytic solutions to simple model problems displaying increasingly complex solution behavior.

References

1. U. Ascher and R. Weiss, "Collocation for singular perturbation problems II: Linear first order systems without turning points,"

Technical Report 82-4, Department of Computer Science, University of British Columbia, Vancouver, 1982.

2. N. Fenichel, "Geometric singular perturbation theory for ordinary differential equations," J. Differential Equations 31 (1979), 53-98.

3. J. E. Flaherty and R. E. O'Malley, Jr., "Asymptotic and numerical methods for vector systems of singularly perturbed boundary value problems," Proceedings, Army Numerical Analysis and Computer Conference, Vicksburg, 1982.

4. M. I. Freedman and B. Granoff, "Formal asymptotic solution of a singularly perturbed nonlinear optimal control problem," J. Optim. Theory Appl. 19 (1976), 301-325.

5. M. I. Freedman and J. L. Kaplan, "Singular perturbations of two-point boundary value problems arising in optimal control," SIAM J. Control Optim. 14 (1976), 189-215.

6. M. I. Freedman and J. L. Kaplan, "Perturbation analysis of an optimal control problem involving bang-bang controls," J. Differential Equations 25 (1977), 11-29.

7. V. Ja. Glizer and M. G. Dmitriev, "Singular perturbations in a linear optimal control problem with quadratic functional," Soviet Math Dokl. 16 (1975), 1555-1558.

8. V. Ja. Glizer and M. G. Dmitriev, "Singular perturbations and generalized functions," Soviet Math. Dokl. 20 (1979), 1360-1364.

9. C. R. Hadlock, "Existence and dependence on a parameter of solutions of a nonlinear two point boundary value problem," J. Differential Equations 14 (1973), 498-517.

10. W. A. Harris, Jr., "Singularly perturbed boundary value problems revisited," Lecture Notes in Math. 312, Springer-Verlag, Berlin, 1973, 54-64.

11. F. Hoppensteadt, "Properties of solutions of ordinary differential equations with a small parameter," Comm. Pure Appl. Math. 24 (1971), 807-840.

12. F. A. Howes, "Boundary-interior layer interactions in nonlinear singular perturbation theory," Memoirs Amer. Math. Society 203, 1978.

13. P. V. Kokotovic and P. Sannuti, "Singular perturbation method for reducing the model order in optimal control design," IEEE Trans. Automatic Control 13 (1968), 377-384.

14. H. L. Kurland, "Solutions to boundary value problems of fast-slow systems by continuing homology in the Morse index along a path of isolated invariant sets of the fast systems," preprint, Boston University, 1981.

15. J. J. Levin and N. Levinson, "Singular perturbations of nonlinear systems and an associated boundary layer equation," J. Rational Mech. Anal. 3 (1954), 247-270.

16. H. D. McIntyre, "The formal asymptotic solution of a nonlinear singularly perturbed state regulator," pre-thesis monograph, University of Arizona, 1977.

17. R. E. O'Malley, Jr., "Singular perturbation of a boundary value problem for a system of nonlinear differential equations," J. Differential Equations 8 (1970), 431-447.

18. R. E. O'Malley, Jr., Introduction to Singular Perturbations, Academic Press, New York, 1974.

19. R. E. O'Malley, Jr., "Singular perturbations and optimal control," Lecture Notes in Math. 680, Springer-Verlag, Berlin, 1978, 170-218.

20. R. E. O'Malley, Jr., "On multiple solutions of singularly perturbed systems in the conditionally stable case," Singular Perturbations and Asymptotics, R. E. Meyer and S. V. Parter, editors, Academic Press, New York, 1980, 87-108.

21. R. E. O'Malley, Jr., "Slow/fast decoupling -- Analytical and numerical aspects," preprint, Rensselaer Polytechnic Institute, 1982.

22. P. Sannuti, "Asymptotic expansions of singularly perturbed quasi-linear optimal systems," SIAM J. Control 13 (1975), 572-592.

23. P. Sannuti and P. Kokotovic, "Singular perturbation method for near optimal design of high-order non-linear systems," Automatica 5 (1969), 773-779.

24. S. S. Sastry, C. A. Desoer and P. P. Varaiya, "Jump behavior of circuits and systems," Memorandum UCB/ERL M80/44, University of California, Berkeley, 1980.

25. R. Weiss, "An analysis of the box and trapezoidal schemes for linear singularly perturbed boundary value problems," preprint, Technische Universität Wein, 1982.

# SINGULAR PERTURBATIONS IN NONLINEAR SYSTEMS AND OPTIMAL CONTROL

P. HABETS

Institut de Mathématique
Université Catholique de Louvain
Chemin du Cyclotron, 2
B-1348 LOUVAIN-LA-NEUVE (Belgium).

## INTRODUCTION - THE NONLINEAR REGULATOR PROBLEM -

### 1. Problem statement.

Let us consider a nonlinear system

$$\begin{aligned} \dot{x} &= f(t,x,y,u,\varepsilon), \quad x(o) = x_o(\varepsilon), \\ \varepsilon\dot{y} &= g(t,x,y,u,\varepsilon), \quad y(o) = y_o(\varepsilon), \end{aligned} \tag{0.1}$$

where $x \in \mathbb{R}^n$, $y \in \mathbb{R}^m$ are state vectors, $u \in \mathbb{R}^r$ is a control vector, $t \in [0,1]$ and $\varepsilon > 0$ is a small parameter. The objective is to find a control $u(t)$ which minimizes the cost functional

$$J = \pi(x(1),y(1),\varepsilon) + \int_0^1 V(t,x(t),y(t),u(t),\varepsilon)dt. \tag{0.2}$$

It is well known (cfr Athans & Falb [1]) that necessary conditions for an optimal control can be obtained from the Hamiltonian

$$H = -V(t,x,y,u,\varepsilon) + p^T f(t,x,y,u,\varepsilon) + q^T g(t,x,y,u,\varepsilon).$$

Here $p \in \mathbb{R}^n$ and $\varepsilon q \in \mathbb{R}^m$ are the costate variables associated with x and y and the superscript T denotes transpose. The Pontryagin maximum principle implies that the optimal trajectory, costate and control are such that

$$\begin{aligned}
\dot{x} &= f(t,x,y,u,\varepsilon), \quad x(o) = x_o, \\
\dot{p} &= -\frac{\partial H}{\partial x}(t,x,p,y,q,u,\varepsilon), \quad p(1) = -\frac{\partial \pi}{\partial x}(x(1),y(1),\varepsilon), \\
\varepsilon\dot{y} &= g(t,x,y,u,\varepsilon), \quad y(o) = y_o, \\
\varepsilon\dot{q} &= -\frac{\partial H}{\partial y}(t,x,p,y,q,u,\varepsilon), \quad q(1) = -\frac{1}{\varepsilon}\frac{\partial \pi}{\partial y}(x(1),y(1),\varepsilon), \\
0 &= -\frac{\partial H}{\partial u}(t,x,p,y,q,u,\varepsilon).
\end{aligned} \tag{0.3}$$

Our aim is to compute solutions of this two point boundary value problem (BVP) for small values of $\varepsilon$. The classical procedure is two steps. First, using experience, intuition or any other trick, one computes a function

$$w^*(t,\varepsilon) = (x^*,p^*,y^*,q^*,u^*),$$

called a *formal solution*, that satisfies "almost" the equations (0.3) (e.g. with error terms of order $O(\varepsilon^N)$). Next, one proves the existence of a solution

$$w(t,\varepsilon) = (x,p,y,q,u)$$

of (0.3) near $w^*(t,\varepsilon)$ (e.g. such that

$$w(t,\varepsilon) = w^*(t,\varepsilon) + O(\varepsilon^N)).$$

The control problem (0.1) (0.2) was already studied by Kokotović & Sannuti [2](1968).

## 2. The linear problem.

The simplest situation occurs when f,g are linear, V is quadratic and

$$\pi = \frac{1}{2}(x^T(1),y^T(1)) \begin{pmatrix} \pi_1 & \varepsilon\pi_2 \\ \varepsilon\pi_2^T & \varepsilon\pi_3 \end{pmatrix} \begin{pmatrix} x(1) \\ y(1) \end{pmatrix}.$$

In this case the BVP (0.3) is linear and the boundary values are bounded as $\varepsilon \to 0$. O'Malley [3](1972) gives sufficient conditions to compute a formal solution in a decoupled form

$$w^{*}(t,\varepsilon) = w^{0}(t,\varepsilon) + w^{L}(\frac{t}{\varepsilon},\varepsilon) + w^{R}(\frac{1-t}{\varepsilon},\varepsilon), \qquad (0.4)$$

where

$$w^{0}(t,\varepsilon) = \Sigma\, w_i^{o}(t)\varepsilon^{i},$$

$$w^{L}(\tau,\varepsilon) = \Sigma\, w_i^{L}(\tau)\varepsilon^{i} = O(e^{-\mu\tau}), \qquad (0.5)$$

$$w^{R}(\sigma,\varepsilon) = \Sigma\, w_i^{R}(\sigma)\varepsilon^{i} = O(e^{-\mu\sigma}), \qquad (0.6)$$

and $\mu > 0$. This structure of the formal solution implies that on any closed subinterval of (0,1)

$$w^{*}(t,\varepsilon) = w^{0}(t,\varepsilon) + O(\varepsilon^{N}).$$

For this reason, $w^{0}(t,\varepsilon)$ is called the *outer solution* (i.e. away from the set {0,1}). On the other hand,

$$w^{L}(\frac{t}{\varepsilon},\varepsilon) \text{ and } w^{R}(\frac{1-t}{\varepsilon},\varepsilon)$$

describe rapid changes of $w^{*}(t,\varepsilon)$ in a neighbourhood of $t = 0$ and $t = 1$. They represent what is called the *boundary layers*. We shall see that this decoupling of the outer solution from the boundary layers is due to the fact that they satisfy exponential bounds (0.5) (0.6).

In the constant coefficient case, it is easy to relate this decoupling with the geometry of eigenspaces. The outer solution $w^{0}(t,\varepsilon)$ lies in a "slow" eigenspace corresponding to eigenvalues which remain bounded as $\varepsilon \to 0$. The left boundary layer $w^{L}(\frac{t}{\varepsilon},\varepsilon)$ belongs to a "fast" eigenspace corresponding to eigenvalues $\lambda(\varepsilon)$ such that

$$\text{Re } \lambda(\varepsilon) \to -\infty \text{ as } \varepsilon \to 0.$$

Similarly, the right boundary layer $w^{R}(\frac{1-t}{\varepsilon},\varepsilon)$ lies in a "fast" eigenspace with eigenvalues $\lambda(\varepsilon)$ such that

$$\text{Re } \lambda(\varepsilon) \to +\infty \text{ as } \varepsilon \to 0.$$

3. A quasilinear problem.

The structure (0.4) of the formal solution generalizes to a larger class of systems. Sannuti [4](1974) and O'Malley [5](1974) gave sufficient conditions for such a formal solution to exist when the control problem is such that (0.3) is linear in y,u but nonlinear in x. Such a problem is called *quasilinear*. Sannuti [6](1975) justifies this formal approach in case

$$J = \int_0^1 [V(t,x,(t)) + u^T(t)R(t)u(t)]dt.$$

4. Nonlinear problem.

In order to justify a first order formal solution, Hadlock [7](1973) considers the BVP

$$\begin{aligned} \dot{x} &= f(t,x,p,y,q), \quad x(o) = x_0, \\ \dot{p} &= F(t,x,p,y,q), \quad p(1) = p_1, \\ \varepsilon\dot{y} &= g(t,x,p,y,q), \quad y(o) = y_0, \\ \varepsilon\dot{q} &= G(t,x,p,y,q), \quad q(1) = q_1. \end{aligned} \tag{0.7}$$

He shows that if

$$w^*(t) = (x^*,p^*,y^*,q^*)$$

satisfies the reduced BVP

$$\begin{aligned} \dot{x} &= f(t,x,p,y,q), \quad x(o) = x_0, \\ \dot{p} &= F(t,x,p,y,q), \quad p(1) = p_1, \\ 0 &= g(t,x,p,y,q), \\ 0 &= G(t,x,p,y,q), \end{aligned}$$

and if the boundary layer's jump

$$|y_0 - y^*(o)|, \quad |q_1 - q^*(1)|$$

are small enough, then

$$(x,p) = (x^*(t),p^*(t)) + \sigma(1)$$

$$(y,q) = (y^*(t),q^*(t)) + \sigma(1) + O(e^{-\mu t/\varepsilon} + e^{-\mu(1-t)/\varepsilon}).$$

This justifies the first order formal solution $w^*$ provided the constraint

$$\frac{\partial H}{\partial u}(t,x,y,p,q,u) = 0$$

can be solved for the control u.

A $n^{th}$ order formal solution is investigated in Freedman & Granoff [8] (1976) and a justification is given in Freedman & Kaplan [9] (1976) under the same restriction on the boundary layer jump.

In all these papers, the boundary conditions are as in (0.7). This supposes

$$\pi(x(1),y(1)) = -p_1x(1) - \varepsilon q_1 y(1). \tag{0.8}$$

Vasil'eva & Anikeeva [10] (1976) consider a case where

$$\pi(x(1),y(1)) = -p_1x(1) - q_1y(1).$$

This implies

$$q(1) = \frac{q_1}{\varepsilon}$$

and the solution is unbounded as $\varepsilon \to 0$. An other extension can be found in O'Malley [11] (1980). He considers the formal solutions of a BVP with nonlinear boundary conditions.

5. In the first part of the present paper, we shall investigate formal solutions of the BVP

$$\begin{aligned} \dot{x} &= f(t,x,y,u), \quad Px(o) + (I-P)x(1) = \alpha, \\ \varepsilon\dot{y} &= g(t,x,y,u), \quad Qy(o) + (I-Q)y(1) = \beta, \\ 0 &= h(t,x,y,u) \end{aligned} \tag{0.9}$$

where $x,f,\alpha \in \mathbb{R}^n$, $y,g,\beta \in \mathbb{R}^m$, $u,h \in \mathbb{R}^r$, $t \in [0,1]$ $\varepsilon > 0$ and $P,Q$ are projection matrices

$$P = \begin{bmatrix} I_l & 0 \\ 0 & 0 \end{bmatrix}, \quad Q = \begin{bmatrix} I_k & 0 \\ 0 & 0 \end{bmatrix},$$

$(0 \leqslant l \leqslant n,\ 0 \leqslant k \leqslant m)$. This applies to the control problem (0.1) (0.2) in case (0.8) is satisfied.

In the second and third parts, we shall justify the procedure in two different cases. The first one, the quasilinear case, makes use of a structural assumption on (0.9). In the second case, the key assumption is somewhat of a restriction on the size of the boundary layer's jump.

This generalizes and unifies previous approaches.

At last, let us notice that we can assume the BVP (0.9) to depend explicitly on $\varepsilon$. The case where the boundary conditions are nonlinear needs more attention and will be published elsewhere.

PART ONE - THE FORMAL SOLUTION -

## 1. Preliminary results.

Consider the BVP (0.9) and let us compute a $N^{th}$ order formal solution. More precisely, for each $N > 0$, we look for a function $w^*(t,\varepsilon) = (x^*,y^*,u^*)$ such that

$$\begin{aligned} \dot{x}^* &= f(t,x^*,y^*,u^*) + O(\varepsilon^N), \quad Px^*(0,\varepsilon) + (I-P)x^*(1,\varepsilon) = \alpha + O(\varepsilon^N) \\ \varepsilon\dot{y}^* &= g(t,x^*,y^*,u^*) + O(\varepsilon^N), \quad Qy^*(0,\varepsilon) + (I-Q)y^*(1,\varepsilon) = \beta + O(\varepsilon^N) \\ 0 &= h(t,x^*,y^*,u^*) + O(\varepsilon^N). \end{aligned}$$

Motivated by previous work (e.g. by the linear case), we want $w^*$ to be of the form (0.4). The determination of such a function $w^*$ will be based on the following lemmas.

**LEMMA 1.** (Decoupling)

*Assume* $f : \mathbb{R} \times \mathbb{R}^s \to \mathbb{R}$, $(t,w) \to f(t,w)$ *is a* $C^2$ *function and let*

$$w = w(t,\varepsilon) = w^0(t,\varepsilon) + w^L(\frac{t}{\varepsilon},\varepsilon) + w^R(\frac{1-t}{\varepsilon},\varepsilon)$$

*be such that*

$$w^0(t,\varepsilon) = O(1), \quad w^L(\tau,\varepsilon) = O(e^{-\mu\tau}), \quad w^R(\sigma,\varepsilon) = O(e^{-\mu\sigma}).$$

*Then one can write*

$$\begin{aligned} f(t,w(t,\varepsilon)) &= f(t,w^0(t,\varepsilon)) \\ &+ f(t,w^0(t,\varepsilon) + w^L(\frac{t}{\varepsilon},\varepsilon)) - f(t,w^0(t,\varepsilon)) \\ &+ f(t,w^0(t,\varepsilon) + w^R(\frac{1-t}{\varepsilon},\varepsilon)) - f(t,w^0(t,\varepsilon)) \\ &+ R(t,\varepsilon), \end{aligned}$$

*where*

$$R(t,\varepsilon) = O(e^{-\frac{\mu}{\varepsilon}}).$$

**Remark.**

It follows that for any $N \in \mathbb{N}$, $R(t,\varepsilon) = O(\varepsilon^N)$.

Proof.

One verifies that for some $\lambda$ and $\lambda_i \in (0,1)$, $i=1,\dots,s$,

$$R = [\, f(t,w^0 + w^L + w^R) - f(t,w^0+w^L)\,] - [\, f(t,w^0+w^R) - f(t,w^0)\,]$$

$$= [\frac{\partial f}{\partial w}(t,w^0 + \lambda w^L + w^R) - \frac{\partial f}{\partial w}(t,w^0 + \lambda w^L)\,]\, w^L$$

and

$$\frac{\partial f}{\partial w_i}(t,w^0 + \lambda w^L + w^R) - \frac{\partial f}{\partial w_i}(t,w^0 + \lambda w^L) = \frac{\partial^2 f}{\partial w\, \partial w_i}(t,w^0 + \lambda w^L + \lambda_i w^R) w^R =$$

$$= O(e^{-\frac{\mu(1-t)}{\varepsilon}}).$$

Hence

$$R = O(e^{-\frac{\mu(1-t)}{\varepsilon}})\, w^L = O(e^{-\frac{\mu(1-t)}{\varepsilon}} e^{-\frac{\mu t}{\varepsilon}}) = O(e^{-\frac{\mu}{\varepsilon}}). \qquad \square$$

**Remark.**

Lemma 1 decouples the function f into three terms that correspond respectively to the outer solution, the left and the right boundary layers. For such a decomposition to be possible, one needs

$$R(t,\varepsilon) = O(\varepsilon^N)$$

i.e.

$$\| w^L \| . \| w^R \| = O(\varepsilon^N).$$

This means that the boundary layers cannot interfer with each other up to the order $\varepsilon^N$.

LEMMA 2. (Outer expansion of f)

*Assume* $f : \mathbb{R} \times \mathbb{R}^s \to \mathbb{R}$, $(t,w) \to f(t,w)$ *is a* $C^{N+1}$ *function and let*

$$w(t,\varepsilon) = \sum_0^N w_i(t)\varepsilon^i$$

*be such that*

$$w_i(t) = O(1).$$

*Then one can write*

$$f(t,w(t,\varepsilon)) = f(t,w_0(t)) + \sum_1^N [\frac{\partial f}{\partial w}(t,w_0(t))w_k(t) + f_k^0(t,w_0,\dots,w_{k-1})\,]\varepsilon^k + O(\varepsilon^{N+1}), \tag{1.1}$$

*where* $f_k^0 = O(1)$.

**Proof.**

From Taylor's expansion, one has

$$f(t,w(t,\varepsilon)) = f(t,w_0) + \sum_1^N \frac{1}{k!} d^k f(t,w_0;\varepsilon w_1+\dots+\varepsilon^N w_N) + R(t,\varepsilon),$$

where

$$d^k f(t,w_0;h) = \sum_{i_1,\dots,i_k=1}^{s} \frac{\partial^{i_1+\dots+i_k} f}{\partial w_{i_1}\dots\partial w_{i_k}}(t,w_0)h_{i_1}\dots h_{i_k}$$

and for some $\lambda \in (0,1)$

$$R(t,\varepsilon) = \frac{1}{(N+1)!} d^{N+1} f(t,w_0 + \lambda(w-w_0);w - w_0) = O(\varepsilon^{N+1}).$$

Regrouping terms of equal power in $\varepsilon$, one obtains (1.1). □

The next lemmas take care of the boundary layers. It will be essential in the proof that they satisfy exponential bounds.

**LEMMA 3.** (Inner expansion of f)

*Assume* $f : \mathbb{R}^{1+s} \to \mathbb{R}$, $(t,w) \to f(t,w)$ *is a* $C^{N+1}$ *function and let*

$$w^0(t,\varepsilon) = \sum_0^N w_i(t)\varepsilon^i, \quad w^L(\tau,\varepsilon) = \sum_0^N w_i^L(\tau)\varepsilon^i$$

*be such that*

$$w_i(t) = O(1), \quad w_i^L(\tau) = O(e^{-\mu\tau}).$$

*Then one can write*

$$\begin{aligned} f(t,w^0(t,\varepsilon) &+ w^L(\tfrac{t}{\varepsilon},\varepsilon)) - f(t,w^0(t,\varepsilon)) = \\ &= f(0,w_0^0(0) + w_0^L(\tfrac{t}{\varepsilon})) - f(0,w_0^0(0)) \\ &+ \sum_1^N \varepsilon^k \left[\frac{\partial f}{\partial w}(0,w_0^0(0) + w_0^L(\tfrac{t}{\varepsilon}))w_k^L(\tfrac{t}{\varepsilon}) + f_k^L(\tfrac{t}{\varepsilon},w_0^L,\dots,w_{k-1}^L)\right] \\ &+ O(\varepsilon^{N+1}), \end{aligned} \tag{1.2}$$

*where each functions* $f_k^L$ *is bounded.*

**Remark.**

Each function $f_k^L(\tau,w_0,\dots,w_{k-1})$ is a polynomial in $\tau,w_1,\dots,w_{k-1}$ with coefficients function of $w_0$.

**Proof.**

Let us introduce the function

$$\Delta(t,\varepsilon,w) = f(t,w^0(t,\varepsilon) + w) - f(t,w^0(t,\varepsilon)). \qquad (1.3)$$

From Taylor's expansion, one has for some $\lambda \in (0,1)$

$$\Delta(t,\varepsilon,w^L) = \Delta(0,0,w_0^L) + \sum_1^N \frac{1}{k!} d^k\Delta(0,0,w_0^L;t,\varepsilon,w^L - w_0^L) + \qquad (1.4)$$
$$\frac{1}{(N+1)!} d^{N+1}\Delta(\lambda t,\lambda\varepsilon,w_0^L + \lambda(w^L - w_0^L);\ t,\varepsilon,w^L - w_0^L),$$

where

$$d^k\Delta\ (s,\eta,v;t,\varepsilon,w) =$$
$$= \sum_{\alpha+\beta+\gamma_1+\dots+\gamma_s=k} \left[\frac{\partial^k\Delta}{\partial t^\alpha\ \partial\varepsilon^\beta\ \partial w_1^{\gamma_1}\dots\partial w_s^{\gamma_s}}(s,\eta,v)\right]\left(\frac{t}{\varepsilon}\right)^\alpha \left(\frac{w_1}{\varepsilon}\right)^{\gamma_1}\dots\left(\frac{w_s}{\varepsilon}\right)^{\gamma_s}\varepsilon^k$$

and

$$w = (w_1,\dots,w_s)$$

Next we write

$$w^L = \sum_0^N w_i^L\left(\frac{t}{\varepsilon}\right)\varepsilon^i \quad \text{and} \quad w_i^L = w_i^L\left(\frac{t}{\varepsilon}\right) = (w_{i1}^L,\dots,w_{is}^L),\ i=0,\dots,N.$$

We can expand

$$\left(\frac{(w^L - w_0^L)_j}{\varepsilon}\right)^{\gamma_j} = (w_{1j}^L + \varepsilon\, w_{2j}^L + \dots + \varepsilon^{N-1}w_{Nj}^L)^{\gamma_j}$$
$$= \sum A_{\gamma_j^1\dots\gamma_j^N}(w_{1j}^L)^{\gamma_j^1}\dots(w_{Nj}^L)^{\gamma_j^N}\varepsilon^{\gamma_j^2 + 2\gamma_j^3 + \dots (N-1)\gamma_j^N} \qquad (1.5)$$

If we introduce these expressions in (1.4) and collect terms of equal power in $\varepsilon$, one obtains

$$\Delta(t,\varepsilon,w^L) = \Delta(0,0,w_0^L)$$
$$+ \sum_{k=1}^N \left[\frac{\partial\Delta}{\partial w}(0,0,w_0^L)\, w_k^L + \Delta_k^L\left(\frac{t}{\varepsilon},w_0^L,\dots,w_{k-1}^L\right)\right]\varepsilon^k$$
$$+ R(t,\varepsilon). \qquad (1.6)$$

From (1.4) and (1.5), it is clear that the functions $\Delta_k^L$ are linear **combinations** of terms of the form

$$C = \left[\frac{\partial^{\alpha+\beta+\gamma_1+\dots+\gamma_s}\Delta}{\partial t^\alpha\ \partial\varepsilon^\beta\ \partial w_1^{\gamma_1}\dots\partial w_s^{\gamma_s}}(0,0,w_0^L)\right]\left(\frac{t}{\varepsilon}\right)^\alpha \prod_{i,j}(w_{ij}^L)^{\gamma_j^i}.$$

If $\gamma_1 + \dots + \gamma_s \neq 0$, one has

$$C = O\left(\left(\frac{t}{\varepsilon}\right)^{\alpha} \left(e^{-\frac{\mu t}{\varepsilon}}\right)^{\Sigma\gamma_j^i}\right) = O(1).$$

On the other hand, if $\gamma_1 + \dots + \gamma_s = 0$, one has for some $\overline{\lambda} \in (0,1)$

$$C = \left[\frac{\partial^{\alpha+\beta}\Delta}{\partial t^{\alpha}\partial\varepsilon^{\beta}}(0,0,w_0^L)\right]\left(\frac{t}{\varepsilon}\right)^{\alpha}$$

$$= \left[\frac{\partial^{\alpha+\beta}}{\partial t^{\alpha}\partial\varepsilon^{\beta}}\left[f(t,w^0(t,\varepsilon)+w) - f(t,w^0(t,\varepsilon)\right]_{(0,0,w_0^L)}\right]\left(\frac{t}{\varepsilon}\right)^{\alpha}$$

$$= \left[\frac{\partial^{\alpha+\beta}}{\partial t^{\alpha}\partial\varepsilon^{\beta}}\left[\frac{\partial}{\partial w} f(t,w^0(t,\varepsilon)+\overline{\lambda}w)w\right]_{(0,0,w_0^L)}\right]\left(\frac{t}{\varepsilon}\right)^{\alpha}$$

$$= O\left(\|w^L\|\left(\frac{t}{\varepsilon}\right)^{\alpha}\right) = O\left(\left(\frac{t}{\varepsilon}\right)^{\alpha} e^{-\frac{\mu t}{\varepsilon}}\right) = O(1).$$

Hence, each function $\Delta_k^L$ is bounded. Using the same argument, one proves that

$$\left[\frac{\partial^{\alpha+\beta+\gamma_1+\dots+\gamma_s}\Delta}{\partial t^{\alpha}\,\partial\varepsilon^{\beta}\,\partial w_1^{\gamma_1}\dots\partial w_s^{\gamma_s}}(\lambda t,\lambda\varepsilon,w_0^L+\lambda(w^L-w_0^L))\right]\left(\frac{t}{\varepsilon}\right)^{\alpha}\prod_{ij}(w_{ij}^L)^{\gamma_j^i} = O(1).$$

Consequently

$$R(t,\varepsilon) = O(\varepsilon^{N+1}). \tag{1.7}$$

At last, from (1.3) (1.6) and (1.7), one deduces (1.2). □

<u>LEMMA 4</u>.

*Assume* $f : \mathbb{R}^{1+s} \to \mathbb{R}$, $(t,w) \to f(t,w)$ *is a* $C^{N+1}$ *function and let*

$$w^0(t,\varepsilon) = \sum_0^N w_i(t)\varepsilon^i,\quad w^R(\sigma,\varepsilon) = \sum_0^N w_i^R(\sigma)\varepsilon^i$$

*be such that*

$$w_i(t) = O(1),\ w_i^R(\sigma) = O(e^{-\mu\sigma}).$$

*Then one can write*

$$f(t,w^o(t,\varepsilon) + w^R(\frac{1-t}{\varepsilon},\varepsilon)) - f(t,w^o(t,\varepsilon)) =$$

$$= f(1,w_o^o(1) + w_o^R(\frac{1-t}{\varepsilon})) - f(1,w_o^o(1))$$

$$+ \sum_1^N \varepsilon^k[\frac{\partial f}{\partial w}(1,w_o^o(1)+w_o^R(\frac{1-t}{\varepsilon}))w_k^R(\frac{1-t}{\varepsilon}) + f_k^R(\frac{1-t}{\varepsilon},w_o^R,\dots,w_{k-1}^R)]$$

$$+ O(\varepsilon^{N+1})$$

*where each function* $f_k^R$ *is bounded.*

2. Formal expansions of solutions.

As mentioned above, we shall compute a formal solution $w^*$ of (0.9) such that

$$w^*(t,\varepsilon) = \sum_o^N w_i^o(t)\varepsilon^i + \sum_o^N w_i^L(\frac{t}{\varepsilon})\varepsilon^i + \sum_o^N w_i^R(\frac{1-t}{\varepsilon})\varepsilon^i.$$

If we assume

$$w_i^o(t) = O(1),\ w_i^L(\tau) = O(e^{-\mu\tau}) \text{ and } w_i^R(\sigma) = O(e^{-\mu\tau}),$$

one computes from lemmas 1, 2, 3 and 4

$$\dot{x}^*(t,\varepsilon) - f(t,w^*(t,\varepsilon)) =$$

$$\sum_o^N \dot{x}_i^o(t)\varepsilon^i + \sum_o^N \frac{d}{d\tau}x_i^L(\frac{t}{\varepsilon})\varepsilon^{i-1} - \sum_o^N \frac{d}{d\sigma}x_i^R(\frac{1-t}{\varepsilon})\varepsilon^{i-1}$$

$$- \{f(t,w_o^o(t)) + \sum_1^N \varepsilon^k[\frac{\partial f}{\partial w}(t,w_o^o(t))w_k(t) + f_k^o(t,w_o^o,\dots,w_{k-1}^o)]$$

$$+ f(o,w_o^o(o)+w_o^L(\frac{t}{\varepsilon})) - f(o,w_o^o(o))$$

$$+ \sum_1^N \varepsilon^k[\frac{\partial f}{\partial w}(o,w_o^o(o)+w_o^L(\frac{t}{\varepsilon}))w_k^L(\frac{t}{\varepsilon}) + f_k^L(\frac{t}{\varepsilon},w_o^L,\dots,w_{k-1}^L)]$$

$$+ f(1,w_o^o(1)+w_o^R(\frac{1-t}{\varepsilon})) - f(1,w_o^o(1))$$

$$+ \sum_{1}^{N} \varepsilon^k [\frac{\partial f}{\partial w}(1, w_o^o(1) + w_o^R(\frac{1-t}{\varepsilon})) w_k^R(\frac{1-t}{\varepsilon}) + f_k^R(\frac{1-t}{\varepsilon}, w_o^R, \ldots, w_{k-1}^R)]$$

$$+ O(\varepsilon^{N+1})\}.$$

Hence

$$\dot{x}^*(t,\varepsilon) = f(t, w^*(t,\varepsilon)) + O(\varepsilon^{N+1}),$$

if we choose the coefficients $w_i^o$, $w_i^L$ and $w_i^R$ such that

$$\dot{x}_o^o(t) = f(t, w_o^o(t)),$$

$$\dot{x}_i^o(t) = \frac{\partial f}{\partial w}(t, w_o^o(t)) w_i(t) + f_i^o(t, w_o^o, \ldots, w_{i-1}^o),$$

$$\frac{d}{d\tau} x_o^L(\tau) = 0,$$

$$\frac{d}{d\tau} x_1^L(\tau) = f(o, w_o^o(o) + w_o^L(\tau)) - f(o, w_o^o(o)),$$

$$\frac{d}{d\tau} x_{i+1}^L(\tau) = \frac{\partial f}{\partial w}(o, w_o^o(o) + w_o^L(\tau)) w_i^L(\tau) + f_i^L(\tau, w_o^L, \ldots, w_{i-1}^L),$$

$$\frac{d}{d\sigma} x_o^R(\sigma) = 0,$$

$$\frac{d}{d\sigma} x_1^R(\sigma) = - f(1, w_o^o(1) + w_o^R(\sigma)) + f(1, w_o^o(1)),$$

$$\frac{d}{d\sigma} x_{i+1}^R(\sigma) = \frac{\partial f}{\partial w}(1, w_o^o(1) + w_o^R(\sigma)) w_i^R(\sigma) + f_i^R(\sigma, w_o^R, \ldots, w_{i-1}^R).$$

Similarly, one can write

$$Px^*(o,\varepsilon) + (1-P)x^*(1,\varepsilon) - \alpha =$$

$$= \Sigma(Px_k^o(o) + (I-P)x_k^o(1))\varepsilon^k$$

$$+ \Sigma\, Px_k^L(o)\varepsilon^k + \Sigma(I-P)x_k^R(o)\varepsilon^k + O(\varepsilon^{N+1}) - \alpha$$

$$= O(\varepsilon^{N+1})$$

if

$$Px_o^o(o) + (I-P)x_o^o(1) = \alpha - Px_o^L(o) - (I-P)x_o^R(o),$$

$$Px_i^o(o) + (I-P)x_i^o(1) = - Px_i^L(o) - (I-P)x_i^R(o), \; i=2,\ldots,N.$$

If we write similar equations corresponding to the other equations of (2.1) and collect terms of equal power in ε, we obtain determining equations for the functions

$$w_i^O(t),\ w_i^L(\tau),\ w_i^R(\sigma).$$

The *outer solution* $\Sigma\, w_i^O(t)\varepsilon^i$ should satisfy the equations

$$O_o\ \begin{cases} \dot{x}_o^O = f(t,w_o^O(t)),\ Px_o^O(o) + (I-P)x_o^O(1) = \alpha - Px_o^L(o) - (I-P)x_o^R(o),\\ 0 = g(t,w_o^O(t)),\\ 0 = h(t,w_o^O(t)),\end{cases}$$

$$O_k\ \begin{cases} \dot{x}_k^O = \frac{\partial f}{\partial w}(t,w_o^O(t))w_k + f_k^O(t,w_o^O,\dots,w_{k-1}^O),\\ 0 = \frac{\partial g}{\partial w}(t,w_o^O(t))w_k + g_k^O(t,w_o^O,\dots,w_{k-1}^O) - \dot{y}_{k-1}^O,\\ 0 = \frac{\partial h}{\partial w}(t,w_o^O(t))w_k + h_k^O(t,w_o^O,\dots,w_{k-1}^O),\\ \qquad Px_k^O(o) + (I-P)x_k^O(1) = -Px_k^L(o) - (I-P)x_k^R(o).\end{cases}$$

The *left inner solution* $\Sigma\, w_i^L(\tau)\varepsilon^i$ satisfies

$$L_{-1}\quad \dot{x}_o^L(\tau) = 0,$$

$$L_o\ \begin{cases} \dot{x}_1^L(\tau) = f(o,w_o^O(o) + w_o^L(\tau)) - f(o,w_o^O(o)),\\ \dot{y}_o^L(\tau) = g(o,w_o^O(o) + w_o^L(\tau)) - g(o,w_o^O(o)),\ Qy_o^L(o) = Q\beta - Qy_k^O(o),\\ 0 = h(o,w_o^O(o) + w_o^L(\tau)) - h(o,w_o^O(o)),\end{cases}$$

$$L_k\ \begin{cases} \dot{x}_{k+1}^L(\tau) = \frac{\partial f}{\partial w}(o,w_o^O(o) + w_o^L(\tau))w_k^L(\tau) + f_k^L(\tau,w_o^L,\dots,w_{k-1}^L),\\ \dot{y}_k^L(\tau) = \frac{\partial g}{\partial w}(o,w_o^O(o) + w_o^L(\tau))w_k^L(\tau) + g_k^L(\tau,w_o^L,\dots,w_{k-1}^L),\ Qy_k^L(o) = -Qy_k^O(o),\\ 0 = \frac{\partial h}{\partial w}(o,w_o^O(o) + w_o^L(\tau))w_k^L(\tau) + h_k^L(\tau,w_o^L,\dots,w_{k-1}^L).\end{cases}$$

The *right inner solution* $\Sigma\, w_i^R(\sigma)\varepsilon^i$ satisfies

$$R_{-1} \quad \dot{x}_0^R(\sigma) = 0,$$

$$R_0 \quad \begin{cases} \dot{x}_1^R(\sigma) = -f(1,w_0^0(1)+w_0^R(\sigma)) + f(1,w_0^0(1)), \\ \dot{y}_0^R(\sigma) = -g(1,w_0^0(1)+w_0^R(\sigma)) + g(1,w_0^0(1)), \\ 0 = -h(1,w_0^0(1)+w_0^R(\sigma)) + h(1,w_0^0(1)), \\ \qquad (I-Q)y_0^R(o) = (I-Q)\beta - (I-Q)y_0^0(1), \end{cases}$$

$$R_k \quad \begin{cases} \dot{x}_{k+1}^R(\sigma) = -[\frac{\partial f}{\partial w}(1,w_0^0(1)+w_0^R(\sigma))w_k^R(\sigma) + f_k^R(\sigma,w_0^R,\dots,w_{k-1}^R)], \\ \dot{y}_k^R(\sigma) = -[\frac{\partial g}{\partial w}(1,w_0^0(1)+w_0^R(\sigma))w_k^R(\sigma) + g_k^R(\sigma,w_0^R,\dots,w_{k-1}^R)], \\ 0 = -[\frac{\partial h}{\partial w}(1,w_0^0(1)+w_0^R(\sigma))w_k^R(\sigma) + h_k^R(\sigma,w_0^R,\dots,w_{k-1}^R)], \\ \qquad (I-Q)y_k^R(o) = -(I-Q)y_k^0(1). \end{cases}$$

These computations lead to the following expansion algorithm.

PROPOSITION 5.

*Assume that the systems*

$$L_{-1},\ R_{-1},\ O_0,\ L_0,\ R_0,\ \dots,\ O_k,\ L_k,\ R_k,\ \dots,\ O_N,\ L_N,\ R_N$$

*can be solved successively by functions*

$$w_k^0(t) = O(1),\quad w_k^L(\tau) = O(e^{-\mu\tau}),\quad w_k^R(\sigma) = O(e^{-\mu\sigma}).$$

*Then the function*

$$w^*(t,\varepsilon) = \sum_{k=o}^{N+1} (w_k^0(t) + w_k^L(\frac{t}{\varepsilon}) + w_k^R(\frac{1-t}{\varepsilon}))\varepsilon^k$$

*is such that*

$$\dot{x}^*(t,\varepsilon) = f(t,w^*(t,\varepsilon)) + O(\varepsilon^{N+1}),\quad Px^*(o,\varepsilon) + (I-P)x^*(1,\varepsilon) = \alpha + O(\varepsilon^{N+1}),$$
$$\varepsilon\dot{y}^*(t,\varepsilon) = g(t,w^*(t,\varepsilon)) + O(\varepsilon^{N+1}),\quad Qy^*(o,\varepsilon) + (I-Q)y^*(1,\varepsilon) = \beta + O(\varepsilon^{N+1}),$$
$$0 = h(t,w^*(t,\varepsilon)) + O(\varepsilon^{N+1}).$$

Notice that the only solution of $L_{-1}$ such that $x_o^L = O(e^{-\mu\tau})$ is $x_o^L \equiv 0$. Similarly, one has $x_o^R \equiv 0$.

**Remark 1.**

Let us choose the functions

$$w_i^o,\ i=0,\ \ldots,\ N\ ;\ w_j^L,\ w_j^R,\ j=0,\ \ldots,\ N-1\ ;\ x_N^L,\ x_N^R,$$

as in proposition 5, and

$$w_N^L = O(e^{-\mu\tau}),\ w_N^R = O(e^{-\mu\sigma})$$

such that

$$Qy_N^L(o) = -\ Qy_N^o(o),\ (I-Q)y_N^R = -\ (I-Q)y_N^o(1).$$

The corresponding function $w^*(t,\varepsilon)$ verifies

$$\begin{aligned}\dot{x}^* - f(t,w^*) &= \varepsilon^N\ [\frac{\partial f}{\partial w}(o,w_o^o(o) + w_o^L(\frac{t}{\varepsilon}))w_N^L(\frac{t}{\varepsilon}) + f^L] \\ &+ \varepsilon^N\ [\frac{\partial f}{\partial w}(1,w_o^o(1) + w_o^R(\frac{1-t}{\varepsilon}))w_N^R(\frac{1-t}{\varepsilon}) + f_N^R] \\ &+ O(\varepsilon^{N+1}) \\ &= O(\varepsilon^N(\varepsilon+\hat{\psi})),\end{aligned}$$

where

$$\hat{\psi} = e^{-\mu t/\varepsilon} + e^{-\mu(1-t)/\varepsilon}.$$

Similarly, one computes

$$\begin{aligned}\varepsilon\dot{y}^* - g(t,w^*) &= O(\varepsilon^N(\varepsilon+\hat{\psi})), \\ h(t,w^*) &= O(\varepsilon^N(\varepsilon+\hat{\psi})),\end{aligned}$$

and

$$\begin{aligned}Px^*(o,\varepsilon) + (I-P)x^*(1,\varepsilon) &= \alpha + O(\varepsilon^{N+1}), \\ Qy^*(o,\varepsilon) + (I-Q)y^*(1,\varepsilon) &= \beta + O(\varepsilon^{N+1}).\end{aligned}$$

A function $w^*$ that satisfies such error estimates can be proved to be a good approximation of a solution of (0.9).

**Remark 2.**

If we choose the functions

$$w_i^o,\ i=0,\ \ldots,\ N\ ;\ w_j^L,\ w_j^R,\ j=0,\ \ldots,\ N-1\ ;\ x_N^L,\ x_N^R,$$

as in remark 1 and

$$w_N^L = 0,\ w_N^R = 0,$$

the corresponding function $w^*(t,\varepsilon)$ satisfies similar equations with boundary conditions

$$Px^*(o,\varepsilon) + (I-P)x^*(1,\varepsilon) = \alpha + O(\varepsilon^{N+1})$$
$$Qy^*(o,\varepsilon) + (I-Q)y^*(1,\varepsilon) = \beta + O(\varepsilon^{N}).$$

PART TWO - THE QUASILINEAR CASE -

## 1. Assumptions.

In order to justify the formal work of part one, we shall need several assumptions that we list here for convenience.

**ASSUMPTION A1.**

*There exists a formal solution*

$$w^*(t,\varepsilon) = (x^*(t,\varepsilon), y^*(t,\varepsilon), u^*(t,\varepsilon))$$

*with graph in a bounded domain and such that*

$$\begin{aligned} \dot{x}^* &= f(t,x^*,y^*,u^*) + O(\eta(\varepsilon+\hat{\psi})), \quad Px^*(0,\varepsilon) + (I-P)x^*(1,\varepsilon) = \alpha + O(\varepsilon\eta), \\ \varepsilon\dot{y}^* &= g(t,x^*,y^*,u^*) + O(\eta(\varepsilon+\hat{\psi})), \quad Qy^*(0,\varepsilon) + (I-Q)y^*(1,\varepsilon) = \beta + O(\eta), \\ 0 &= h(t,x^*,y^*,u^*) + O(\eta(\varepsilon+\hat{\psi})), \end{aligned}$$

*where*

$$\hat{\psi} = \hat{\psi}(t,\varepsilon) = e^{-2\mu t/\varepsilon} + e^{-2\mu(1-t)/\varepsilon}, \quad \eta = O(1), \ \mu > 0.$$

Notice that the formal solutions computed in part one satisfy assumption A1 with $\eta = \varepsilon^N$. In particular remark 2 applies. The case of a reduced solution considered in Hadlock [7] corresponds to $\eta = 1$.

**ASSUMPTION A2.**

*The functions* f, g, h *are* $C^2$.

**ASSUMPTION A3.**

*For some* $\delta > 0$, *the matrices*

$$(1+\varepsilon^{-1}\psi)^{-1}\frac{\partial^2(f,g,h)}{\partial x^2}, \quad \frac{\partial^2(f,g,h)}{\partial x\partial y}, \quad \frac{\partial^2(f,g,h)}{\partial x\partial u},$$

$$(1+\varepsilon^{-1}\psi)\frac{\partial^2(f,g,h)}{\partial y^2}, \quad (1+\varepsilon^{-1}\psi)\frac{\partial^2(f,g,h)}{\partial y\partial u},$$

$$(1+\varepsilon^{-1}\psi)\frac{\partial^2(f,g,h)}{\partial u^2}$$

*are uniformly bounded on the set* D *of points* (t,x,y,u) *such that*

$$t \in [0,1],\ |x-x^*| \leqslant \delta,\ |y-y^*|,\ |u-u^*| \leqslant \delta(1+\varepsilon^{-1}\psi)$$

*where*

$$\psi = \psi(t,\varepsilon) = e^{-\mu t/\varepsilon} + e^{-\mu(1-t)/\varepsilon}.$$

In case f, g, h are linear in y and u, assumption A3 is a consequence of A2. Suppose for instance

$$f = f_1(x) + f_2(x)y + f_3(x)u.$$

From A2, the matrices

$$\frac{\partial^2 f_1}{\partial x^2},\ \frac{\partial^2 f_2}{\partial x^2},\ \frac{\partial^2 f_3}{\partial x^2}$$

are uniformly bounded in the set of points x such that

$$|x-x^*| \leqslant \delta.$$

It follows that on D

$$\begin{aligned}(1+\varepsilon^{-1}\psi)^{-1}\left|\frac{\partial^2 f}{\partial x^2}\right| &\leqslant \left|\frac{\partial^2 f_1}{\partial x^2}\right| + \left|\frac{\partial^2 f_2}{\partial x^2}\right|\left(|y^*| + (1+\varepsilon^{-1}\psi)^{-1}|y-y^*|\right)\\ &\quad + \left|\frac{\partial^2 f_3}{\partial x^2}\right|\left(|u^*| + (1+\varepsilon^{-1}\psi)^{-1}|u-u^*|\right)\\ &\leqslant K.\end{aligned}$$

One has also

$$\left|\frac{\partial^2 f}{\partial x \partial y}\right| = \left|\frac{\partial f_2}{\partial x}\right| \leqslant K$$

and

$$\left|\frac{\partial^2 f}{\partial x \partial u}\right| = \left|\frac{\partial f_3}{\partial x}\right| \leqslant K.$$

Using similar relations for g and h, one sees that assumption A3 generalizes the situation investigated in Sannuti [6]. For this reason we call it the *quasilinear case.*

**ASSUMPTION A4.**

*The matrix*

$$H_3 = \frac{\partial h}{\partial u}(t,w^*)$$

*is regular.*

ASSUMPTION A5.

*The matrix*

$$\overline{G}_2 = \left(\frac{\partial g}{\partial y} - \frac{\partial g}{\partial u}\left(\frac{\partial h}{\partial u}\right)^{-1} \frac{\partial h}{\partial y}\right)(t,w^*)$$

*has* k *eigenvalues with real part* $\leqslant -2\mu < 0$ *and* m-k *eigenvalues with real part* $\geqslant 2\mu > 0$.

ASSUMPTION A6.

*If* $Y(t)$ *is a fundamental matrix of*

$$\varepsilon\dot{y} = \overline{G}_2 y, \tag{2.1}$$

*one has*

$$Y(t)(QY(o) + (I-Q)Y(1))^{-1} = O(\varepsilon+\psi).$$

It is known that assumption A5 implies the existence of a $C^1$-matrix $R(t,\varepsilon)$ such that

$$R^{-1}\overline{G}_2 R = \begin{pmatrix} G_- & 0 \\ 0 & G_+ \end{pmatrix} \tag{2.2}$$

where $G_-$ has k eigenvalues with negative real part and $G_+$ has m-k eigenvalues with positive real part.

If $w^* = w^o(t)$ is a first order outer solution, the matrix R is independent of $\varepsilon$. Let us partition R according to (2.2)

$$R = \begin{pmatrix} R_1 & R_2 \\ R_3 & R_4 \end{pmatrix}.$$

PROPOSITION 6.

*Suppose the matrix* R *defined in* (2.2) *is independent of* $\varepsilon$ *and the matrices*

$$R_1(o),\ R_4(1)$$

*are regular. Then assumption* A6 *is satisfied.*

**Proof.**

The change of variables

$$y = R(t)v$$

is such that (2.1) becomes

$$\varepsilon\dot{v} = \begin{pmatrix} G_- & \\ & G_+ \end{pmatrix} v - \varepsilon R^{-1}\dot{R}v. \tag{2.3}$$

From assumption A5 and Lemma 1 in Chang & Coppel [12], the system

$$\varepsilon\dot{v}_- = G_-v_-$$

has a fundamental matrix $V_-(t)$ such that

$$V_-(0) = I, \quad |V_-(t)V_-^{-1}(s)| \leq Ke^{-\mu(t-s)/\varepsilon}, \quad 1 \geq t \geq s \geq 0.$$

Similarly, there exists a fundamental matrix $V_+(t)$ of

$$\varepsilon\dot{v}_+ = G_+v_+$$

such that

$$V_+(1) = I, \quad |V_+(t)V_+^{-1}(s)| \leq Ke^{-\mu(s-t)/\varepsilon}, \quad 1 \geq s \geq t \geq 0.$$

Consider the Banach space $\mathcal{B}$ of continuous matrix valued functions $V(t)$, $t \in [0,1]$ with norm

$$\|V\| = \max_t |V(t)|.$$

A fundamental matrix of (2.3) is a fixed point of the operator

$$T : \mathcal{B} \to \mathcal{B}, \quad V \mapsto T(V)$$

where

$$T(V)(t) = \begin{pmatrix} V_-(t) & \\ & V_+(t) \end{pmatrix} - \int_0^t \begin{bmatrix} V_-(t)V_-^{-1}(s) & 0 \\ 0 & 0 \end{bmatrix} R^{-1}(s)\dot{R}(s)V(s)ds$$
$$+ \int_t^1 \begin{bmatrix} 0 & 0 \\ 0 & V_+(t)V_+^{-1}(s) \end{bmatrix} R^{-1}(s)\dot{R}(s)V(s)ds.$$

One verifies that

$$\|T(V_1)-T(V_2)\| \leqslant K \sup\left(\int_0^t |V_-(t)V_-^{-1}(s)|ds + \int_t^1 |V_+(t)V_+^{-1}(s)|ds\right) \|V_1-V_2\|$$
$$\leqslant K\varepsilon \|V_1-V_2\|.$$

Hence, from Banach's fixed point theorem

$$V(t) = \begin{pmatrix} V_-(t) & \\ & V_+(t) \end{pmatrix} + O(\varepsilon)$$

and

$$Y(t) = R(t) \begin{pmatrix} V_-(t) & \\ & V_+(t) \end{pmatrix} + O(\varepsilon).$$

It follows

$$QY(o) + (I-Q)Y(1) = QR(o)Q + (I-Q)R(1)(I-Q) + O(\varepsilon)$$

and

$$Y(t)(QY(o) + (I-Q)Y(1))^{-1} =$$
$$= R(t) \begin{pmatrix} V_-(t) & \\ & V_+(t) \end{pmatrix} \begin{pmatrix} R_1^{-1}(o) & \\ & R_4^{-1}(1) \end{pmatrix} + O(\varepsilon)$$
$$= O(\varepsilon+\psi).$$ □

**ASSUMPTION A7.**

*If* $X(t)$ *is a fundamental matrix of*

$$\dot{x} = \tilde{F}_1 x, \tag{2.4}$$

*where*

$$\tilde{F}_1 = \bar{F}_1 - \bar{F}_2\bar{G}_2^{-1}\bar{G}_1$$

$$\bar{F}_1 = \left(\frac{\partial f}{\partial x} - \frac{\partial f}{\partial u}\left(\frac{\partial h}{\partial u}\right)^{-1} \frac{\partial h}{\partial x}\right)(t,w^*)$$

$$\bar{F}_2 = \left(\frac{\partial f}{\partial y} - \frac{\partial f}{\partial u}\left(\frac{\partial h}{\partial u}\right)^{-1} \frac{\partial h}{\partial y}\right)(t,w^*)$$

and

$$\bar{G}_1 = \left(\frac{\partial g}{\partial x} - \frac{\partial g}{\partial u}\left(\frac{\partial h}{\partial u}\right)^{-1} \frac{\partial h}{\partial x}\right)(t,w^*),$$

*the matrix*

$$PX(o) + (I-P)X(1)$$

*has a bounded inverse.*

The equation (2.4) is equivalent to the linearized reduced problem

$$\dot{x} = \frac{\partial f}{\partial x} x + \frac{\partial f}{\partial y} y + \frac{\partial f}{\partial u} u$$

$$0 = \frac{\partial g}{\partial x} x + \frac{\partial g}{\partial y} y + \frac{\partial g}{\partial u} u$$

$$0 = \frac{\partial h}{\partial x} x + \frac{\partial h}{\partial y} y + \frac{\partial g}{\partial u} u.$$

Notice also that if $\tilde{F}_1$ is independent of $\varepsilon$, assumption A7 is equivalent to assume the BVP

$$\dot{x} = \tilde{F}_1 x$$

$$Px(o) + (I-P)x(1) = 0$$

has only the zero solution.

**ASSUMPTION A8.**

*The matrix*

$$\bar{G}_2^{-1}\bar{G}_1$$

*is such that*

$$\frac{d}{dt} \bar{G}_2^{-1}\bar{G}_1 = O(1+\varepsilon^{-1}\hat{\psi}).$$

If $w^*$ is of the form (0.4), A8 follows from A2 and regularity assumptions on $w^*$.

## 2. The variational equations.

Let us introduce the variables

$$\bar{x} = x-x^*, \ \bar{y} = y-y^*, \ \bar{u} = u-u^*$$

and the notation

$$\bar{w} = (\bar{x},\bar{y},\bar{u}).$$

The system (0.9) becomes

$$\begin{aligned} \dot{\bar{x}} &= F_1\bar{x} + F_2\bar{y} + F_3\bar{u} + F(t,\bar{w},\varepsilon), \\ \varepsilon\dot{\bar{y}} &= G_1\bar{x} + G_2\bar{y} + G_3\bar{u} + G(t,\bar{w},\varepsilon), \\ 0 &= H_1\bar{x} + H_2\bar{y} + H_3\bar{u} + H(t,\bar{w},\varepsilon), \end{aligned} \tag{2.5}$$

and

$$\begin{aligned} P\bar{x}(o) + (I-P)\bar{x}(1) &= \bar{\alpha}, \\ Q\bar{y}(o) + (I-Q)\bar{y}(1) &= \bar{\beta}, \end{aligned} \tag{2.6}$$

where

$$\begin{pmatrix} F_1 & F_2 & F_3 \\ G_1 & G_2 & G_3 \\ H_1 & H_2 & H_3 \end{pmatrix} = \frac{\partial(f,g,h)}{\partial(x,y,u)}(t,w^*),$$

$$F(t,\bar{w},\varepsilon) = f(t,w^*+\bar{w}) - f(t,w^*) - \frac{\partial f}{\partial w}(t,w^*)\bar{w} + O(\eta(\varepsilon+\hat{\psi})),$$

$$G(t,\bar{w},\varepsilon) = g(t,w^*+\bar{w}) - g(t,w^*) - \frac{\partial g}{\partial w}(t,w^*)\bar{w} + O(\eta(\varepsilon+\hat{\psi})),$$

$$H(t,\bar{w},\varepsilon) = h(t,w^*+\bar{w}) - h(t,w^*) - \frac{\partial h}{\partial w}(t,w^*)\bar{w} + O(\eta(\varepsilon+\hat{\psi})),$$

$$\bar{\alpha} = O(\varepsilon\eta),\ \bar{\beta} = O(\eta).$$

From Taylor's formula, we can write for each component of F

$$\begin{aligned} F_i(t,\bar{w},\varepsilon) &= \frac{1}{2}\,\bar{w}^T\,\frac{\partial^2 f_i}{\partial w^2}\,(t,w^*+\theta_i\bar{w})\bar{w} + O(\eta(\varepsilon+\hat{\psi})),\ 0 < \theta_i < 1, \\ &= (\bar{x},(1+\varepsilon^{-1}\psi)^{-1}\bar{y},(1+\varepsilon^{-1}\psi)^{-1}\bar{u})\ A \begin{pmatrix} \bar{x} \\ (1+\varepsilon^{-1}\psi)^{-1}\bar{y} \\ (1+\varepsilon^{-1}\psi)^{-1}\bar{u} \end{pmatrix} + O(\eta(\varepsilon+\hat{\psi})), \end{aligned}$$

where

$$A = \frac{1+\varepsilon^{-1}\psi}{2} \begin{pmatrix} (1+\varepsilon^{-1}\psi)^{-1}\dfrac{\partial^2 f_i}{\partial x^2} & \dfrac{\partial^2 f_i}{\partial x\partial y} & \dfrac{\partial^2 f_i}{\partial x\partial u} \\ \dfrac{\partial^2 f_i}{\partial x\partial y} & (1+\varepsilon^{-1}\psi)\dfrac{\partial^2 f_i}{\partial y^2} & (1+\varepsilon^{-1}\psi)\dfrac{\partial^2 f_i}{\partial y\partial u} \\ \dfrac{\partial^2 f_i}{\partial x\partial u} & (1+\varepsilon^{-1}\psi)\dfrac{\partial^2 f_i}{\partial y\partial u} & (1+\varepsilon^{-1}\psi)\dfrac{\partial^2 f_i}{\partial u^2} \end{pmatrix},$$

and assumption A3 implies

$$F_i(t,\bar{w},\varepsilon) = O(|\bar{w}|_\varepsilon^2(1+\varepsilon^{-1}\psi) + \eta(\varepsilon+\hat{\psi}))$$

with

$$|\bar{w}|_\varepsilon = \max(|\bar{x}|,\ (1+\varepsilon^{-1}\psi)^{-1}|\bar{y}|,\ (1+\varepsilon^{-1}\psi)^{-1}|u|).$$

In a similar way, one sees that

$$F,G,H = O(|w|_\varepsilon^2(1+\varepsilon^{-1}\psi) + \eta(\varepsilon+\hat{\psi})). \tag{2.7}$$

Notice also that

$$|H(t,\bar{w}_1,\varepsilon) - H(t,\bar{w}_2,\varepsilon)| \leqslant K(|\bar{w}_1|_\varepsilon + |\bar{w}_2|_\varepsilon)|\bar{w}_1-\bar{w}_2|_\varepsilon(1+\varepsilon^{-1}\psi). \tag{2.8}$$

From assumption A4, the matrix $H_3$ is regular and we can introduce the variable

$$\tilde{u} = H_3^{-1}H_1\bar{x} + H_3^{-1}H_2\bar{y} + \bar{u}. \tag{2.9}$$

System (2.5) becomes

$$\begin{aligned} \dot{\bar{x}} &= \bar{F}_1\bar{x} + \bar{F}_2\bar{y} + \bar{F}(t,\bar{x},\bar{y},\tilde{u},\varepsilon), \\ \varepsilon\dot{\bar{y}} &= \bar{G}_1\bar{x} + \bar{G}_2\bar{y} + \bar{G}(t,\bar{x},\bar{y},\tilde{u},\varepsilon), \end{aligned} \tag{2.10}$$

$$\tilde{u} = -H_3^{-1}H, \tag{2.11}$$

where

$$\begin{pmatrix} \bar{F}_1 & \bar{F}_2 \\ \bar{G}_1 & \bar{G}_2 \end{pmatrix} = \begin{pmatrix} F_1 - F_3H_3^{-1}H_1 & F_2 - F_3H_3^{-1}H_2 \\ G_1 - G_3H_3^{-1}H_1 & G_2 - G_3H_3^{-1}H_2 \end{pmatrix}$$

and

$$\bar{F} = F - F_3H_3^{-1}H,$$

$$\bar{G} = G - G_3H_3^{-1}H.$$

## 3. Diagonalization of the linear part of the equations.

To investigate the linear part of the equations we shall need the following lemma due to Chang[13].

**LEMMA 7.**

*There exists* $\varepsilon_0 > 0$ *such that for any* $\varepsilon \in [0,\varepsilon_0]$ *the equations*

$$\varepsilon\dot{T} = \bar{G}_2 T - \varepsilon T\bar{F}_1 + \varepsilon T\bar{F}_2 T - \bar{G}_1,$$

$$\varepsilon\dot{S} = \varepsilon(\bar{F}_1 - \bar{F}_2 T)S - S(\bar{G}_2 + \varepsilon T\bar{F}_2) - \bar{F}_2,$$

*have solutions*

$$T = T(t,\varepsilon) = \bar{G}_2^{-1}\bar{G}_1 + O(\varepsilon+\psi), \tag{2.12}$$

$$S = S(t,\varepsilon) = O(1).$$

Notice that the stability assumption A5 is essential in order to obtain the existence of bounded solutions T and S. The estimate (2.12) is obtained from assumption A8 and integration by parts.

Consider the change of variables

$$\begin{pmatrix}\tilde{x}\\ \tilde{y}\end{pmatrix} = \begin{pmatrix} I+\varepsilon ST & \varepsilon S\\ T & I\end{pmatrix}\begin{pmatrix}\bar{x}\\ \bar{y}\end{pmatrix}, \quad \begin{pmatrix}\bar{x}\\ \bar{y}\end{pmatrix} = \begin{pmatrix} I & -\varepsilon S\\ -T & I+\varepsilon ST\end{pmatrix}\begin{pmatrix}\tilde{x}\\ \tilde{y}\end{pmatrix}. \tag{2.13}$$

From (2.10) and lemma 7, one computes

$$\begin{pmatrix}\dot{\tilde{x}}\\ \varepsilon\dot{\tilde{y}}\end{pmatrix} = \begin{pmatrix}\bar{F}_1 - \bar{F}_2\bar{G}_2^{-1}\bar{G}_1 & 0\\ 0 & \bar{G}_2\end{pmatrix}\begin{pmatrix}\tilde{x}\\ \tilde{y}\end{pmatrix} + \begin{pmatrix}\hat{F}_1\tilde{x} + \tilde{F}\\ \hat{G}_1\tilde{y} + \tilde{G}\end{pmatrix} \tag{2.14}$$

where

$$\hat{F}_1 = \bar{F}_2(\bar{G}_2^{-1}\bar{G}_1 - T) = O(\varepsilon+\psi)$$

$$\hat{G}_1 = \varepsilon T\bar{F}_2 = O(\varepsilon)$$

$$\begin{pmatrix}\tilde{F}\\ \tilde{G}\end{pmatrix} = \begin{pmatrix} I+\varepsilon ST & S\\ \varepsilon T & I\end{pmatrix}\begin{pmatrix}F - F_3H_3^{-1}H\\ G - G_3H_3^{-1}H\end{pmatrix}.$$

Notice that the change of variables (2.11) (2.13) is such that

$$|\bar{x}| \leqslant K(|\tilde{x}| + \varepsilon|\tilde{y}|),$$

$$|\bar{y}| \leqslant K(|\tilde{x}| + |\tilde{y}|),$$

$$|\bar{u}| \leqslant K(|\tilde{x}| + |\tilde{y}| + |\tilde{u}|).$$

Hence, one has

$$|\overline{w}|_\varepsilon = \max(|\overline{x}|,\ (1+\varepsilon^{-1}\psi)^{-1}|\overline{y}|,\ (1+\varepsilon^{-1}\psi)^{-1}|\overline{u}|)$$

$$\leqslant K \max(|\tilde{x}|,\ (1+\varepsilon^{-1}\psi)^{-1}|\tilde{y}|,\ (1+\varepsilon^{-1}\psi)^{-1}|\tilde{u}|) = K|\tilde{w}|_\varepsilon$$

and from (2.7) (2.8), one computes

$$\tilde{F},\tilde{G} = O(|F| + |G| + |H|) = O(|\tilde{w}|^2_\varepsilon(1+\varepsilon^{-1}\psi) + \eta(\varepsilon+\hat{\psi}))$$

and

$$|H(t,w_1,\varepsilon) - H(t,w_2,\varepsilon)| \leqslant K(|\tilde{w}_1|_\varepsilon + |\tilde{w}_2|_\varepsilon)|\tilde{w}_1-\tilde{w}_2|_\varepsilon(1+\varepsilon^{-1}\psi).$$

The boundary conditions (2.6) are transformed into

$$P\tilde{x}(o) + (I-P)\tilde{x}(1) = \tilde{A}$$
$$Q\tilde{y}(o) + (I-Q)\tilde{y}(1) = \tilde{B} \qquad (2.15)$$

where

$$\tilde{A} = \overline{\alpha} + \varepsilon(PS\tilde{y}(o) + (I-P)S\tilde{y}(1)) = O(\varepsilon|\tilde{y}(o)| + \varepsilon|\tilde{y}(1)| + \varepsilon\eta)$$

and

$$\tilde{B} = \overline{\beta} + QT\overline{x}(o) + (I-Q)T\overline{x}(1) =$$
$$= O(|\tilde{x}(o)| + \varepsilon|\tilde{y}(o)| + |\tilde{x}(1)| + \varepsilon|\tilde{y}(1)| + \eta).$$

Let us notice that refining lemma 7, one can choose T such that $QT(o) + (I-Q)T(1) = 0$. In this case $\tilde{B} = \overline{\beta} = O(\eta)$.

## 4. The fixed point problem.

Consider the Banach space $\mathcal{W}$ of continuous functions

$$w = (x,y,u) : [0,1] \to \mathbb{R}^{n+m+r}$$

together with the norm

$$\|w\| = \max_t |w|_\varepsilon = \max_t(\max(|x|,\ (1+\varepsilon^{-1}\psi)^{-1}|y|,\ (1+\varepsilon^{-1}\psi)^{-1}|u|).$$

To solve the BVP (2.14) (2.11) (2.15) is equivalent to find a fixed point w of the operator

$$\mathcal{T} : \mathcal{W} \to \mathcal{W},\ w \mapsto (\mathcal{T}_1(w),\mathcal{T}_2(w),\mathcal{T}_3(w))$$

where

$$T_1(w)(t) = X(t)x_o(w) + \int_o^t X(t)X^{-1}(s)(\hat{F}_1 x+\tilde{F})_{(s,w(s),\varepsilon)}ds,$$

$$x_o(w) = (PX(o) + (I-P)X(1))^{-1} [\hat{A}(w) - (I-P) \int_o^1 X(1)X^{-1}(s)(\hat{F}_1 x+\tilde{F})ds,$$

$$\hat{A}(w) = \bar{\alpha} + \varepsilon[P(ST_2(w))(o) + (I-P)(ST_2(w))(1)],$$

$$T_2(w)(t) = Y(t)(QY(o) + (I-Q)Y(1))^{-1}y_o(w)$$

$$+ \int_o^t Y(t)QY^{-1}(s)\,\frac{\hat{G}_1 y+\tilde{G}}{\varepsilon}\,ds - \int_t^1 Y(t)(I-Q)Y^{-1}(s)\,\frac{\hat{G}_1 y+\tilde{G}}{\varepsilon}\,ds,$$

$$y_o(w) = \tilde{B} + (QY(o)(I-Q) - (I-Q)\,Y(1)Q) \int_o^1 Y^{-1}(s)\,\frac{\hat{G}_1 y+\tilde{G}}{\varepsilon}\,ds,$$

$$T_3(w)(t) = -\,H_3^{-1}H(t,\bar{w},\varepsilon),$$

and $\bar{w}$ is obtained from w by the transformation (2.9) (2.13)

$$w = \begin{pmatrix} I+\varepsilon ST & \varepsilon S & 0 \\ T & I & 0 \\ H_3^{-1}H_1 & H_3^{-1}H_2 & I \end{pmatrix} \bar{w}.$$

LEMMA 8.

*There exists* $k > 0$ *such that for* $\varepsilon$ *small enough*

$$T(B) \subset B,$$

*where* B *is the set of functions* $w \in C$ *such that*

$$|x(t)| \leq k\eta\varepsilon, \quad |y(t)|, |u(t)| \leq k\eta(\varepsilon+\psi).$$

**Proof.**

Let $w \in B$. Notice that if k is fixed and $\varepsilon$ small enough, $k\eta\varepsilon \leq \delta$ and assumption A3 applies. Next one computes

(a) $|w|_\varepsilon \leq k\eta\varepsilon$;

(b) $$|y_o(w)| \leq K(|\tilde{B}| + \int_o^1 \frac{\psi}{\varepsilon}\,|\hat{G}y + \tilde{G}|ds)$$
$$\leq K(k^2\eta^2\varepsilon + k\eta\varepsilon + \eta)(1 + \int_o^1 \psi(1+\varepsilon^{-1}\psi))$$
$$\leq K(k^2\eta\varepsilon + k\varepsilon + 1)\eta,$$

since from lemma 1 in Chang & Coppel [12], one has

$$|Y(t)QY^{-1}(s)| \leq Ke^{-\mu(t-s)/\varepsilon},\ 0 \leq s \leq t \leq 1,$$

$$|Y(t)(I-Q)Y^{-1}(s)| \leq Ke^{-\mu(s-t)/\varepsilon},\ 0 \leq t \leq s \leq 1,$$

and

$$\int_0^t \psi \leq K\varepsilon,\ \int_0^t \psi^2 \leq K\varepsilon;$$

(c) $|\int_0^t Y(t)QY^{-1}(s)\ \frac{\hat{G}_1 y+\tilde{G}}{\varepsilon}\ ds| \leq$

$$\leq K \int_0^t \frac{e^{-\mu(t-s)/\varepsilon}}{\varepsilon}\ [(k^2\eta^2\varepsilon^2+k\eta\varepsilon^2)(1+\varepsilon^{-1}\psi) + \varepsilon\eta(1+\varepsilon^{-1}\hat{\psi})]$$

$$\leq K[(k^2\eta^2\varepsilon^2+k\eta\varepsilon^2+\varepsilon\eta) + (k^2\eta^2\varepsilon^{1/2}+k\eta\varepsilon^{1/2})\psi+\eta\psi]$$

$$\leq K(k^2\eta\varepsilon^{1/2}+k\varepsilon^{1/2}+1)\eta(\varepsilon+\psi);$$

(d) $|\int_t^1 Y(t)(I-Q)Y^{-1}(s)\ \frac{\hat{G}_1 y+\tilde{G}}{\varepsilon}\ ds| \leq K(k^2\eta\varepsilon^{1/2}+k\varepsilon^{1/2}+1)\eta(\varepsilon+\psi);$

(e) $|T_2(w)(t)| \leq K(\varepsilon+\psi)|y_0(w)| + |\int_0^t Y(t)QY^{-1}(s)\ \frac{\hat{G}y+\tilde{G}}{\varepsilon}|$

$$+ |\int_t^1 Y(t)(I-Q)Y^{-1}(s)\ \frac{\hat{G}y+\tilde{G}}{\varepsilon}|$$

$$\leq K(k^2\eta\varepsilon^{1/2}+k\varepsilon^{1/2}+1)\eta(\varepsilon+\psi); \tag{2.16}$$

(f) $|x_0(w)| \leq K[|\hat{A}(w)| + \int_0^1 [(\varepsilon+\psi)|x| + |w|_\varepsilon^2(1+\varepsilon^{-1}\psi) + \eta(\varepsilon+\hat{\psi})]]$

$$\leq K(k^2\eta\varepsilon^{1/2}+k\varepsilon^{1/2}+1)\varepsilon\eta;$$

(g) $|T_1(w)(t)| \leq K[|x_0(w)| + \int_0^t [(\varepsilon+\psi)|x| + |w|_\varepsilon^2(1+\varepsilon^{-1}\psi) + \eta(\varepsilon+\hat{\psi})]]$

$$\leq K(k^2\eta\varepsilon^{1/2}+k\varepsilon^{1/2}+1)\varepsilon\eta; \tag{2.17}$$

(h) $|T_3(w)(t)| \leqslant K|H(t,\bar{w},\varepsilon)|$

$$\leqslant K(|w|_\varepsilon^2(1+\varepsilon^{-1}\psi) + \eta(\varepsilon+\hat{\psi}))$$

$$\leqslant K(k^2\eta\varepsilon+1)\varepsilon\eta(1+\varepsilon^{-1}\psi). \tag{2.18}$$

At last, from (2.16) (2.17) (2.18), we can choose k large enough and $\varepsilon_0$ small enough such that for any $\varepsilon \in (0,\varepsilon_0)$, one has $k\eta\varepsilon \leqslant \delta$ and

$$|T_1(w)(t)| \leqslant k\varepsilon\eta, \quad |T_2(w)(t)|, \quad |T_3(w)(t)| \leqslant k\eta(\varepsilon+\psi). \qquad \square$$

**COROLLARY 9.**

*Let R and S be defined as*

$$R : W \to W, \ w \mapsto (T_1(w), \ T_2(w), \ 0 \quad )$$
$$S : W \to W, \ w \mapsto (0 \quad , \ 0 \quad , \ T_3(w)).$$

*Then*

$$R(B) + S(B) \subset B$$

(i.e. $\forall u \in B$, $v \in B$ $\quad R(u) + S(v) \in B$).

We are now ready to apply the following fixed point theorem.

**LEMMA 10.** (Krasnosel'ski [14])

*Let* B *be a closed bounded convex set in a Banach space* $W$, *let* $R : B \to W$ *be completely continuous and* $S : B \to W$ *be a contraction. If further*

$$R(B) + S(B) \subset B$$

*the operator* $R + S$ *has a fixed point in* B.

The application of this lemma proves now our main result.

**THEOREM 11.**

*If assumptions* A1 *to* A8 *are satisfied, the* BVP (0.9) *has, for* $\varepsilon$ *small enough, a solution*

$$x = x^*(t,\varepsilon) + O(\eta\varepsilon),$$
$$y = y^*(t,\varepsilon) + O(\eta(\varepsilon+\psi)),$$
$$u = u^*(t,\varepsilon) + O(\eta(\varepsilon+\psi)).$$

PART THREE - THE NONLINEAR CASE -

1. Assumptions.

In this section, we shall justify a formal solution $w^*$ as obtained in proposition 5. The following assumption introduces in such a result the size $\gamma$ of the $N^{th}$ order boundary layer jump.

**ASSUMPTION B1.**

*There exists a formal solution*

$$w^*(t,\varepsilon) = (x^*(t,\varepsilon),\ y^*(t,\varepsilon),\ u^*(t,\varepsilon))$$

*with graph in a bounded domain, such that*

$$\begin{aligned} \dot{x}^* &= f(t,x^*,y^*,u^*) + O(\eta(\varepsilon+\gamma\hat{\psi})),\quad Px^*(o,\varepsilon) + (I-P)x^*(1,\varepsilon) = \alpha + O(\varepsilon\eta), \\ \varepsilon\dot{y}^* &= g(t,x^*,y^*,u^*) + O(\eta(\varepsilon+\gamma\hat{\psi})),\quad Qy^*(o,\varepsilon) + (I-Q)y^*(1,\varepsilon) = \beta + O(\gamma\eta), \\ 0 &= h(t,x^*,y^*,u^*) + O(\eta(\varepsilon+\gamma\hat{\psi})), \end{aligned}$$

*where*

$$\hat{\psi} = e^{-2\mu t/\varepsilon} + e^{-2\mu(1-t)/\varepsilon},\quad \eta=O(1),\ \mu > 0.$$

Notice that in case $\gamma = O(\varepsilon)$, proposition 5 implies B1.

In the rest of this section, we shall assume together with B1, assumptions A2, A4, A5, A6, A7 and A8.

2. The variational equations.

Let us introduce the variables

$$\bar{x} = x-x^*,\ \bar{y} = y-y^*,\ \bar{u} = u-u^*.$$

As above, we obtain the BVP (2.5) (2.6). Further one computes

$$(F,G,H)_{(t,o,\varepsilon)} = O(\eta(\varepsilon+\gamma\hat{\psi}))$$

and, if $w_1$, $w_2$ belong to some bounded set,

$$|(F,G,H)_{(t,w_1,\varepsilon)} - (F,G,H)_{(t,w_2,\varepsilon)}| \leqslant K(|w_1| + |w_2|)|w_1-w_2|.$$

Next, we introduce the variable

$$\tilde{u} = H_3^{-1}H_1\bar{x} + H_3^{-1}H_2\bar{y} + \bar{u}$$

which transforms (2.5) into (2.10) (2.11).

3. Diagonalization of the linear part of the equations.

The change of variables (2.13)

$$\begin{pmatrix} \tilde{x} \\ \\ \tilde{y} \end{pmatrix} = \begin{pmatrix} I+\varepsilon ST & \varepsilon S \\ \\ T & I \end{pmatrix} \begin{pmatrix} \bar{x} \\ \\ \bar{y} \end{pmatrix}$$

transforms (2.10) into (2.14). It is easy to see that

$$(\tilde{F},\tilde{G},H)(t,o,\varepsilon) = O(\eta(\varepsilon+\gamma\hat{\psi}))$$

and if $w_1$, $w_2$ belong to some bounded set

$$|(\tilde{F},\tilde{G},H)_{(t,w_1,\varepsilon)} - (\tilde{F},\tilde{G},H)_{(t,w_2,\varepsilon)}| \leqslant K(|w_1| + |w_2|)|w_1-w_2|.$$

The boundary conditions (2.6) are transformed into

$$P\tilde{x}(o) + (I-P)\tilde{x}(1) = \tilde{A}(\tilde{y}(o),\tilde{y}(1)),$$

$$Q\tilde{y}(o) + (I-Q)\tilde{y}(1) = \tilde{B}(\tilde{x}(o),\tilde{x}(1),\tilde{y}(o),\tilde{y}(1)),$$

where

$$\tilde{A}(y_o,y_1) = O(\varepsilon|y_o| + \varepsilon|y_1| + \varepsilon\eta)$$

$$\tilde{B}(o) = O(\gamma\eta)$$

and

$$|\tilde{A}(y_{01},y_{11}) - \tilde{A}(y_{02},y_{12})| \leqslant K\varepsilon \max(|y_{01}-y_{02}|,|y_{11}-y_{12}|)$$

$$|\tilde{B}(x_{01},x_{11},y_{01},y_{11}) - \tilde{B}(x_{02},x_{12},y_{02},y_{12})| \leqslant$$

$$\leqslant K \max(|x_{01}-x_{02}|,|x_{11}-x_{12}|, \varepsilon|y_{01}-y_{02}|, \varepsilon|y_{11}-y_{12}|).$$

4. The fixed point problem.

Consider the Banach space $\mathcal{W}$ of continuous functions

$$w : [0,1] \to \mathbb{R}^{n+m+r}$$

with the norm

$$\|w\| = \max(\max_t(|x(t)|, (1+\varepsilon^{-\lambda}\psi)^{-1}|y(t)|, (1+\varepsilon^{-\lambda}\psi)^{-1}|u(t)|)),$$

with $\lambda \in (1/2,1)$. We shall apply Banach's fixed point theorem to the operator $T$ defined in II-4. To this end we compute (for $\varepsilon$ and $\gamma$ bounded)

(a) $|y_o(o)| \leqslant K[|\tilde{B}(o)| + \int_o^1 \frac{\psi}{\varepsilon} |\tilde{G}(o)|]$

$\leqslant K[\gamma\eta + \int_o^1 \psi\eta(1+\gamma \frac{\hat{\psi}}{\varepsilon})] \leqslant K\eta(\gamma+\varepsilon);$

(b) $|T_2(o)| \leqslant K[(\varepsilon+\psi)|y_o(o)| + \int_o^t e^{-\mu(t-s)/\varepsilon} \eta(1+\gamma \frac{\hat{\psi}}{\varepsilon})$

$+ \int_t^1 e^{-\mu(s-t)/\varepsilon} \eta(1+\gamma \frac{\hat{\psi}}{\varepsilon})]$

$\leqslant K[(\varepsilon+\psi)\eta(\gamma+\varepsilon) + \eta(\varepsilon+\gamma\psi)]$

$\leqslant K(\varepsilon^{1-\lambda}+\gamma)\eta\varepsilon^{\lambda}(1+\varepsilon^{-\lambda}\psi);$

(c) $\hat{A}(o) = \overline{\alpha} + \varepsilon[P(ST_2(o))(o) + (I-P)(ST_2(o))(1)]$

$= O(\varepsilon\eta);$

(d) $|x_o(o)| \leqslant K[|\hat{A}(o)| + \int_o^1 |\tilde{F}(o)|]$

$\leqslant K[\varepsilon\eta + \int_o^1 \eta(\varepsilon+\gamma\hat{\psi})] \leqslant K\varepsilon\eta;$

(e) $|T_1(o)| \leqslant K[|x_o(o)| + \int_o^t |\tilde{F}(o)|] \leqslant K\varepsilon\eta;$

(f) $|T_3(o)| \leqslant K|H(t,o,\varepsilon)| \leqslant K\eta(\varepsilon+\gamma\psi)$

$\leqslant K(\varepsilon^{1-\lambda}+\gamma)\eta\varepsilon^{\lambda}(1+\varepsilon^{-\lambda}\psi).$

From (b), (e) and (f), it follows

$$\|T(o)\| \leqslant K(\varepsilon^{1-\lambda}+\gamma)\eta\varepsilon^{\lambda}.$$

Consider next functions $w_i$ (i=1,2) such that $\|w_i\| \leqslant k\eta\varepsilon^{\lambda}$ i.e.

$$|x_i(t)| \leqslant k\eta\varepsilon^{\lambda}, \quad |y_i(t)| \leqslant k\eta(\varepsilon^{\lambda}+\psi), \quad |u_i(t)| \leqslant k\eta(\varepsilon^{\lambda}+\psi).$$

One computes

(a') $$|\Delta B| = |\tilde{B}(x_1(0),x_1(1),y_1(0),y_1(1)) - \tilde{B}(x_2(0),x_2(1),y_2(0),y_2(1))|$$
$$\leqslant K\|w_1-w_2\|;$$

(b') $$|(\tilde{F},\tilde{G})_{(t,w_1,\varepsilon)} - (\tilde{F},\tilde{G})_{(t,w_2,\varepsilon)}| \leqslant$$
$$\leqslant Kk\eta(\varepsilon^{\lambda}+\psi)(1+\varepsilon^{-\lambda}\psi)\|w_1-w_2\|$$
$$\leqslant Kk\eta(\varepsilon^{\lambda}+\varepsilon^{-\lambda}\psi^2)\|w_1-w_2\|;$$

(c') $$|y_0(w_1) - y_0(w_2)| \leqslant K[|\Delta B| + \int_0^1 \psi(|y_1-y_2| + \frac{\tilde{G}(w_1)-\tilde{G}(w_2)}{\varepsilon})ds]$$
$$\leqslant K[1 + \int_0^1 (\psi(1+\varepsilon^{-\lambda}\psi) + \psi\, k\eta\varepsilon^{-1}(\varepsilon^{\lambda}+\varepsilon^{-\lambda}\psi^2))ds]\|w_1-w_2\|$$
$$\leqslant K[1+k\eta\varepsilon^{-\lambda}]\|w_1-w_2\|;$$

(d') $$|\int_0^t Y(t)QY^{-1}(s)\, \frac{\hat{G}_1(y_1-y_2) + \tilde{G}(t,w_1,\varepsilon) - \tilde{G}(t,w_2,\varepsilon)}{\varepsilon}\, ds| \leqslant$$
$$\leqslant K \int_0^t e^{-\frac{\mu(t-s)}{\varepsilon}} (1+\varepsilon^{-\lambda}\psi+k\eta\varepsilon^{-1}(\varepsilon^{\lambda}+\varepsilon^{-\lambda}\psi^2))\|w_1-w_2\|$$
$$\leqslant K[\varepsilon+\varepsilon^{-\lambda}(\varepsilon^2+\varepsilon^{\lambda}\psi) + k\eta\varepsilon^{\lambda} + k\eta\varepsilon^{-\lambda}\psi]\|w_1-w_2\|$$
$$\leqslant K(\varepsilon^{\lambda}+k\eta)(1+\varepsilon^{-\lambda}\psi)\|w_1-w_2\|;$$

(e') $$|\int_t^1 Y(t)(I-Q)Y^{-1}(s)\, \frac{\hat{G}_1(y_1-y_2) + \tilde{G}(t,w,\varepsilon) - \tilde{G}(t,w_2,\varepsilon)}{\varepsilon}\, ds| \leqslant$$
$$\leqslant K(\varepsilon^{\lambda}+k\eta)(1+\varepsilon^{-\lambda}\psi)\|w_1-w_2\|;$$

(f') $$|T_2(w_1) - T_2(w_2)| \leqslant$$
$$\leqslant K[(\varepsilon+\psi)(1+k\eta\varepsilon^{-\lambda}) + (\varepsilon^{\lambda}+k\eta)(1+\varepsilon^{-\lambda}\psi)]\|w_1-w_2\|$$

$$\leqslant K(\varepsilon^{1-\lambda}+k\eta)(1+\varepsilon^{-\lambda}\psi)\|w_1-w_2\|$$

(g') $|x_0(w_1) - x_0(w_2)| \leqslant$

$$\leqslant K[|\hat{A}(w_1) - \hat{A}(w_2)| + \int_0^1 ((\varepsilon+\psi)|x_1-x_2| + |\tilde{F}(w_1) - \tilde{F}(w_2)|)ds]$$

$$\leqslant K[\varepsilon^{1-\lambda} + \int_0^1 (\varepsilon+\psi+k\eta\varepsilon^{\lambda}+k\eta\varepsilon^{-\lambda}\psi^2)ds]\|w_1-w_2\|$$

$$\leqslant K(k\eta+1)\varepsilon^{1-\lambda}\|w_1-w_2\|;$$

(h') $|T_1(w_1)(t) - T_1(w_2)(t)| \leqslant$

$$\leqslant K[|x_0(w_1) - x_0(w_2)| + \int_0^t ((\varepsilon+\psi)|x_1-x_2| + |\tilde{F}(w_1) - \tilde{F}(w_2)|)ds]$$

$$\leqslant K(k\eta+1)\varepsilon^{1-\lambda}\|w_1-w_2\|$$

$$\leqslant K(\varepsilon^{1-\lambda}+k\eta)\|w_1-w_2\|;$$

(i') $|T_3(w_1)(t) - T_3(w_2)(t)| \leqslant K|H(t,\bar{w}_1,\varepsilon) - H(t,\bar{w}_2,\varepsilon)|$

$$\leqslant Kk\eta(\varepsilon^{\lambda}+\varepsilon^{-\lambda}\psi^2)\|w_1-w_2\|$$

$$\leqslant K(\varepsilon^{1-\lambda}+k\eta)(1+\varepsilon^{-\lambda}\psi)\|w_1-w_2\|.$$

It follows from (f') (h') and (i') that

$$\|T(w_1) - T(w_2)\| \leqslant K(\varepsilon^{1-\lambda}+k\eta)\|w_1-w_2\|.$$

Let us choose $k$ and $\varepsilon_0'$ small enough such that if $\varepsilon \in (0,\varepsilon_0']$

$$\|T(w_1) - T(w_2)\| \leqslant \frac{1}{2}\|w_1-w_2\|.$$

Next we choose $\gamma_0$ and $\varepsilon_0 \leqslant \varepsilon_0'$ small enough so that for $\gamma \in [0,\gamma_0]$ and $\varepsilon \in (0,\varepsilon_0]$

$$\|T_0\| \leqslant k\eta\varepsilon^{\lambda}/2.$$

Consequently, we can apply Banach's fixed point theorem in $B(0,k\eta\varepsilon^{\lambda})$. This proves the following.

**THEOREM 12.**

*Under the assumptions* B1, A2 *and* A4 *to* A8 *the* BVP (0.9) *has, for* $\varepsilon$ *and* $\gamma$ *small enough, a solution*

$$\begin{aligned} x &= x^{*}(t,\varepsilon) + O(\eta\varepsilon^{\lambda}) \\ y &= y^{*}(t,\varepsilon) + O(\eta(\varepsilon^{\lambda}+\psi)) \\ u &= u^{*}(t,\varepsilon) + O(\eta(\varepsilon^{\lambda}+\psi)), \end{aligned} \tag{3.1}$$

*where* $\lambda \in (1/2,1)$.

5. <u>High order approximations</u>.

Suppose assumption B1 is satisfied with $\gamma=1$ and $\eta=\sigma(1)$. The same computations apply and we can choose k large enough so that

$$\|T(o)\| \leqslant k\eta\varepsilon^{\lambda}/2.$$

Next we choose $\varepsilon$ small enough so that

$$\|T(w_1) - T(w_2)\| \leqslant \frac{1}{2}\|w_1-w_2\|.$$

Hence we have proved the theorem.

**THEOREM 13.**

*If* $\eta=\sigma(1)$, $\gamma=1$ *and the assumptions* B1, A2 *and* A4 *to* A8 *are satisfied, the* BVP (0.9) *has, for* $\varepsilon$ *small enough, a solution* (x,y,u) *satisfing* (3.1).

<u>**Remark 1.**</u>

We can justify a formal solution

$$w_1^{*}(t) = (x^{o}(t),\ y^{o}(t),\ u^{o}(t))$$

using a bootstrap argument. Let us suppose there exists a formal solution

$$w_2^{*}(t,\varepsilon) = (x^{o}(t)+O(\varepsilon),\ y^{o}(t)+O(\varepsilon+\psi),\ u^{o}(t)+O(\varepsilon+\psi))$$

such that theorem 13 applies with $\eta=\varepsilon^{1-\lambda}$. In practice $w_1^{*}$ is a first order outer solution and $w_2^{*}$ is a second order one. Using theorem 13, one obtains a solution of (0.9) such that

$$x = x_2^*(t,\varepsilon) + O(\varepsilon) = x^0(t) + O(\varepsilon)$$

$$y = y_2^*(t,\varepsilon) + O(\varepsilon+\varepsilon^{1-\lambda}\psi) = x^0(t) + O(\varepsilon+\psi)$$

$$u = u_2^*(t,\varepsilon) + O(\varepsilon+\varepsilon^{1-\lambda}\psi) = u^0(t) + O(\varepsilon+\psi).$$

A direct justification of $w_1^*$ from theorem 12 would lead to

$$x = x^0(t) + O(\varepsilon^\lambda)$$

$$y = y^0(t) + O(\varepsilon^\lambda+\psi)$$

$$u = u^0(t) + O(\varepsilon^\lambda+\psi),$$

under the additional assumption that the boundary jump $\gamma$ is small enough.

<u>Remark 2.</u>

It should be clear that we could have used in theorems 11, 12 and 13 Krasnosel'ski's theorem (lemma 10) or Banach's fixed point theorem as well as other fixed point theorems (e.g. an implicit function theorem as used in Freedman & Kaplan [8]). The main difficulties lie in the study of the linear part of the equations (lemma 7) and the choice of the appropriate norm used in $\mathcal{W}$. At last, let us mention that the same difficulties and ideas already appear in the simpler test problem

$$\varepsilon\ddot{x} + f(t,x,\dot{x}) = 0$$

$$x(o) = a,\ x(1) = b,$$

a good account of which can be found in Erdelyi [15].

## REFERENCES

1. Athans, M. and Falb, P.L., *Optimal Control*, Mc Graw-Hill, New York, 1966.
2. Kokotović, P.V. and Sannuti, P., Singular Perturbation Method for Reducing the Model Order in Optimal Control Design, *IEEE Trans. A.C.*, 4, 377, 1968.
3. O'Malley, R.E. Jr., The Singularly Perturbed Linear State Regulator Problem, *SIAM J. Control*, 10, 399, 1972.
4. Sannuti, P., Asymptotic Series Solution of Singularly Perturbed Optimal Control Problems, *Automatica*, 10, 183, 1974.
5. O'Malley, R.E. Jr., Boundary Layer Methods for Certain Nonlinear Singularly Perturbed Optimal Control Problems, *J. Math. Anal. Appl.*, 45, 468, 1974.
6. Sannuti, P., Asymptotic Expansions of Singularly Perturbed Quasi-linear Optimal Systems, *SIAM J. Control*, 13, 572, 1975.
7. Hadlock, C.R., Existence and Dependence on a Parameter of Solutions of a Nonlinear Two-Point Boundary Value Problem, *J. Diff. Eq.*, 14, 498, 1973.
8. Freedman, M.I. and Granoff, B., Formal Asymptotic Solution of a Singularly Perturbed Nonlinear Optimal Control Problem, *J. Opt. Th. Appl.*, 19, 301, 1976.
9. Freedman, M.I. and Kaplan, J., Singular Perturbations of Two-Point Boundary Value Problems Arising in Optimal Control, *SIAM J. Control*, 14, 189, 1976.
10. Vasil'eva, A.B. and Anikeeva, V.A., Asymptotic Expansion of Solutions of Nonlinear Problems with Singular Boundary Conditions, *Differential Equations* 12, 1235, 1976.
11. O'Malley, R.E. Jr., On Multiple Solutions of Singularly Perturbed Systems in the Conditionally Stable Case, in *Singular Perturbations and Asymptotics*, Meyer, R.E. and Parter, S.V., Eds, Academic Press, New York 1980.

12. Chang, K.W. and Coppel, W.A., Singular Perturbations of Initial Value Problem over a Finite Interval, *Arch. Rational Mech. Anal*, 32, 268, 1969.

13. Chang, K.W., Singular Perturbations of a General Boundary Value Problem, *SIAM J. Math. Anal.*, 3, 520, 1972.

14. Krasnosel'ski, M.A., *Topological Methods in the Theory of Nonlinear Integral Equations*, Pergamon Press, 1954.

15. Erdelyi, A., A Case History in Singular Perturbation, in *International Conference on Differential Equations*, Antosiewicz, H.A., Ed., Academic Press, New York 1975.

# SLOW/FAST DECOUPLING -- ANALYTICAL AND NUMERICAL ASPECTS

Robert E. O'Malley, Jr.
Department of Mathematical Sciences
Rensselaer Polytechnic Institute
Troy, New York 12181
U.S.A.

## 1. Introduction

The numerical solution of linear two-point boundary value problems for vector systems of the form

$$\dot{x} = A(t)x + b(t) \tag{1}$$

$$M_0 x(0) + M_1 x(1) = c \tag{2}$$

is a fundamental question of importance in control theory and throughout science (for such singular perturbation problems, see O'Malley (1974) and (1978)). It is naive to think that much progress has been made when one writes down the variation of parameters formula

*Supported in part by the Office of Naval Research under Contract Number N00014-81K-056.

$$x(t) = X(t)k + \int^t X(t)X^{-1}(s)b(s)ds \qquad (3)$$

for a solution, since it, in large part, merely converts the problem to others involving properties and determination of a fundamental matrix $X(t)$ which satisfies the homogeneous matrix system

$$\dot{X} = \mathcal{A}(t)X. \qquad (4)$$

We shall be particularly concerned with linear systems for which fundamental matrices involve coupled slow and fast dynamics, determined by an $n \times n$ dimensional state matrix $\mathcal{A}(t)$ which is assumed to have high order derivatives. The existence of rapidly-varying solution modes requires $\mathcal{A}$ to have large entries, so we find it natural to introduce a small positive scaling parameter $\varepsilon$ and write

$$\mathcal{A}(t) = \frac{1}{\varepsilon} A(t,\varepsilon) = \frac{1}{\varepsilon} A_0(t) + A_1(t) + \varepsilon A_2(t) + \ldots \qquad (5)$$

with coefficient matrices $A_j$ which are smooth functions of $t$. (More complicated dependence on $\varepsilon$ does occur in practice, and should be dealt with elsewhere.) Thus $X(t)$ satisfies

$$\varepsilon\dot{X} = A(t,\varepsilon)X. \qquad (6)$$

If $A_0$ had all its eigenvalues strictly in the left (or the right) half plane, a bounded matrix solution $X$ could readily be obtained numerically using standard stiff equation solvers (cf. Aiken (1982)). The presence of slowly-varying modes, however, implies that $A(t,0) = A_0(t)$ is a singular matrix. Indeed, fast or rapidly-varying modes will have large, $O(\frac{1}{\varepsilon})$, derivatives compared to smooth or slowly-varying modes. We shall use such a distinction to henceforth label functions as fast or slow. As $\varepsilon \to 0$, then, determining $X$ is a "singular singular perturbation problem" (cf. Vasil'eva and Butuzov (1978) and O'Malley and Flaherty (1980)), so the standard singular perturbation theory does not apply. Although we shall not discuss them, we note that generalizations of our linear analysis have succeeded in practice on some nonlinear problems (cf. Chow et al. (1981) and Hetrick et al. (1981)).

## 2. The Transformation

It is common in both the analytical and numerical literature to seek a change of variables to convert (6) to a block-triangular (or -diagonal) form (cf. e.g., Wasow (1978), Gingold and Hsieh (1982), Kreiss and Kreiss (1981), Mattheij and Staarink (1982), Weiss (1982), O'Malley and Anderson (1982) and, especially, Mattheij and O'Malley (1982) on which this presentation is largely dependent). Specifically, we shall set

$$x = T(t,\varepsilon)y \tag{7}$$

where both T and $T^{-1}$ are smooth (slowly-varying) throughout our interval $0 \leq t \leq 1$. Our transformed problem will necessarily be

$$\varepsilon\dot{y} = U(t,\varepsilon)y \tag{8}$$

where T satisfies the linear system

$$\varepsilon\dot{T} = AT - TU \tag{9}$$

Our object is to simultaneously determine a slowly-varying matrix T and, thereby, $U = T^{-1}(AT-\varepsilon\dot{T})$, so that U is block upper-triangular and such that the corresponding fast and slow (or even the slow, rapidly-growing, rapidly-decaying, and rapidly oscillating) dynamics of (8) are decoupled. We shall first seek a matrix $U(t,\varepsilon)$ of the form

$$U = \begin{pmatrix} U_f & U_m \\ 0 & \varepsilon U_s \end{pmatrix} \tag{10}$$

where $U_f$ is $p \times p$ dimensional and nonsingular throughout [0,1] and $U_s$ is $q \times q$ dimensional. When such a result is possible, we'll call the homogeneous system (4) two-time-scale there with p fast and q slow modes (cf. Anderson (1979)). Note that we have left substantial flexibility to select a smooth solution of (9), because no boundary conditions for T are specified. It is convenient to further split

$$y = \begin{pmatrix} y_1 \\ y_2 \end{pmatrix} \tag{11}$$

and

$$T = (R \quad S) \tag{12}$$

after the first p rows and columns, respectively, so that (8) and (9) decouple into the triangular systems

$$\begin{cases} \varepsilon \dot{y}_1 = U_f(t,\varepsilon) y_1 + U_m(t,\varepsilon) y_2 & (13) \\ \dot{y}_2 = U_s(t,\varepsilon) y_2 & (14) \end{cases}$$

and

$$\begin{cases} \varepsilon \dot{S} = AS - \varepsilon S U_s - R U_m & (15) \\ \varepsilon \dot{R} = AR - R U_f & (16) \end{cases}$$

Note that integration of (14) will be straightforward since $U_s$ is smooth. Moreover, (13) is a "regular" singular perturbation problem since $U_f$ is nonsingular (albeit, perhaps, conditionally stable (cf. O'Malley (1980)).

It is now natural to attempt to determine the smooth $n \times p$ dimensional matrix R as a formal power series

$$R(t,\varepsilon) = R_0 + \varepsilon R_1 + \varepsilon^2 R_2 + \ldots \tag{17}$$

in $\varepsilon$. Substituting into (16) and expanding

$$U_f = U_{f0} + \varepsilon U_{f1} + \ldots \tag{18}$$

implies that we must have

$$A_0R_0 - R_0U_{f0} = 0 \tag{19}$$

and, for each $j \geq 1$,

$$A_0R_j - R_j U_{f0} = R_0 U_{fj} + \alpha_{j-1} \tag{20}$$

where the $\alpha_{j-1}(t)$'s are known in terms of preceding coefficients. We must be able to solve these equations successively for $R_j$ and $U_{fj}$ in a computationally efficient manner. It should be no surprise that the Schur and/or eigen-structure of the leading or principal matrix $A_0(t) = A(t,0)$ will be critical in determining the desired fundamental matrix $X(t)$ or its transformation. Decomposing $A_0$ into its Schur form (cf. Golub and Wilkinson (1976) or Klema and Laub (1980)), we have

$$A_0 = Q_0V_0Q_0' \tag{21}$$

where $Q_0(t)$ is an orthogonal matrix (so $Q_0^{-1} = Q_0'$) and $V_0(t)$ is a (nearly) upper triangular matrix (it might have some $2 \times 2$ diagonal blocks when $A_0$ has complex eigenvalues). $Q_0$ and $V_0$ can be obtained numerically through the QR algorithm. We will assume that they are slowly-varying. We consider their accurate numerical determination as overhead necessary and natural to obtain the behavior of $X(t)$. Since $A_0$ and $V_0$ have rank p (corresponding to the p fast modes of (6)), we can assume the last $q = n - p$ rows of $V_0$ to be trivial. Note that (19) will then be satisfied if we let $R_0$ coincide with the first p columns of $Q_0$ and $U_{f0}$ with the upper $p \times p$ minor of $V_0$. Indeed, we shall be able to take

$$T(t,0) = Q_0(t) \equiv (R_0(t) \quad S_0(t)) \tag{22}$$

and

$$U(t,0) = V_0(t) \equiv \begin{pmatrix} U_{f0}(t) & U_{m0}(t) \\ 0 & 0 \end{pmatrix} \tag{23}$$

To this lowest order (in $\varepsilon$), our change of variables is then analogous to a change to modal coordinates (cf. Porter and Crossley (1972)), though it uses the numerically preferable Schur or singular value decomposition rather than the Jordan canonical form. We note that the eigenvalues of $\frac{1}{\varepsilon} U_{f0}$ coincide asymptotically with the large eigenvalues of $A$.

To obtain higher order terms $R_j$ and $U_{fj}$, it is convenient to set

$$R(t,\varepsilon) = Q_0(t) \begin{pmatrix} I_p + \varepsilon P^1(t,\varepsilon) \\ \varepsilon P^2(t,\varepsilon) \end{pmatrix} \tag{24}$$

so we write

$$R_j = Q_0 \begin{pmatrix} P_j^1 \\ P_j^2 \end{pmatrix} \tag{25}$$

for each $j > 0$. Multiplying (20) by $Q_0'$ and using the Schur decomposition of $A_0$, we obtain the triangular system

$$\begin{cases} U_{f0} P_j^1 - P_j^1 U_{f0} + U_{m0} P_j^2 = U_{fj} + \tilde{\alpha}_{j-1}^{(1)} & (26) \\ P_j^2 U_{f0} = \tilde{\alpha}_{j-1}^{(2)} & (27) \end{cases}$$

where the $\tilde{\alpha}_{j-1}^{(i)}$'s are known successively. Since $U_{f0}$ is nonsingular and essentially upper triangular, it is easy to uniquely solve (27) by back-substitution. There then remains the nonhomogeneous Liapunov equation (26) to determine $P_j^1$ and $U_{fj}$. We can select any solution which respects the Fredholm alternative. A convenient choice is

$$P_j^1 = 0 \text{ and } U_{fj} = U_{m0} P_j^2 - \tilde{\alpha}_{j-1}^{(1)} , \tag{28}$$

though we shall later make another choice for systems with both fast-growing and fast-decaying modes.

Having now specified the first p columns of the transformation matrix T and of the transformed system matrix U (at least, as power series), we select their remaining q columns. We first conveniently take

$$S(t,\varepsilon) = S(t,0) = Q_0(t)\begin{pmatrix} 0 \\ I_q \end{pmatrix} \tag{29}$$

The differential system (15) for S then implies that U must satisfy

$$\varepsilon Q_0' \dot{Q}_0 \begin{pmatrix} 0 \\ I_q \end{pmatrix} = Q_0' A Q_0 \begin{pmatrix} 0 \\ I_q \end{pmatrix} - \varepsilon \begin{pmatrix} 0 \\ I_q \end{pmatrix} U_s - Q_0' R U_m \tag{30}$$

We simplify notation by partitioning the coefficient matrices after their first p rows and columns as

$$Q_0' \dot{Q}_0 = \begin{pmatrix} m_{11} & m_{12} \\ m_{21} & m_{22} \end{pmatrix} \quad \text{and} \quad Q_0' A Q_0 = \begin{pmatrix} n_{11} & n_{12} \\ \varepsilon n_{21} & \varepsilon n_{22} \end{pmatrix}.$$

Using (24), (30) then implies that

$$U_m(t,\varepsilon) = (I_p + \varepsilon P^1(t,\varepsilon))^{-1}(n_{12}(t,\varepsilon) - \varepsilon m_{12}(t)) \tag{31}$$

and

$$U_s(t,\varepsilon) = \varepsilon n_{22}(t,\varepsilon) - m_{22}(t) - P^2(t,\varepsilon)U_m(t,\varepsilon) \tag{32}$$

We note that U is now determined with $U_m(t,0)$ now following from the upper right block of $V_0(t)$ (as in (23)). Moreover, the slow mode system matrix $U_s(t,0)$ is a complicated function of the logarithmic derivative $Q_0' \dot{Q}_0$ of the limiting transformation T(t,0) (through $m_{22}$), $P_0^2$ (cf. (27)), and $U_m(t,0)$. The inverse occuring in (31) certainly exists when $\varepsilon$ is small (or when $P^1 \equiv 0$), and its power series expansion can be easily generated termwise. The approach taken is not, then, simply asymptotic. It will be appropriate whenever the indicated inverse exists. Indeed, substantial numerical interest occurs when $\varepsilon$ is small, but non-vanishing (cf. Söderlind (1981)).

Since our transformation is determined by the nearly orthogonal matrix

$$T(t,\varepsilon) = Q_0 \begin{pmatrix} I_p + \varepsilon P^1 & 0 \\ \varepsilon P^2 & I_q \end{pmatrix}, \tag{33}$$

its inverse is explicitly given by

$$T^{-1}(t,\varepsilon) = \begin{pmatrix} (I_p + \varepsilon P^1)^{-1} & 0 \\ -\varepsilon (I_p + \varepsilon P^1)^{-1} P^2 & I_q \end{pmatrix} Q_0' \tag{34}$$

We should be able to easily verify that the truncated series actually provides decoupling to the corresponding formal order, and that there exist analytic transformations having asymptotic expansions coinciding with the formal series generated (see Wasow (1965) and the literature he cites for proofs of such theory). Numerical experiments are being carried out to verify the computational robustness of the procedure.

## 3. The Two-Point Problem for Systems with Fast-Growing and Fast-Decaying Modes

The principal aim of our decoupling method is to be able to numerically integrate difficult boundary value problems. In terms of our transformed y coordinates, the two-point problem (1)-(2) becomes

$$\begin{cases} \varepsilon \dot{y} = U(t,\varepsilon)y + \varepsilon T^{-1}(t,\varepsilon)b(t) & (35) \\ M_0 T(0,\varepsilon)y(0) + M_1 T(1,\varepsilon)y(1) = c & (36) \end{cases}$$

We'll now restrict attention to problems where the principal matrix $A_0(t)$ of (5) has k eigenvalues with strictly positive real parts,

$\ell = p - k$ eigenvalues strictly in the left half plane, and $q = n - p$ zero eigenvalues throughout the interval $0 \le t \le 1$. In doing so, we are eliminating many turning point problems worthy of study, as well as problems with rapidly oscillating modes. We can then assume that $U_{f0}$ has the form

$$U_{f0} = \begin{pmatrix} U_{f0}^{11} & U_{f0}^{12} \\ 0 & U_{f0}^{22} \end{pmatrix} \tag{37}$$

where $U_{f0}^{11}$ has the k unstable eigenvalues and $U_{f0}^{22}$ the stable ones. In the critical first order, then, we have separated the rapidly growing and rapidly decaying modes (cf. (13)-(14) which only separated fast and slow modes). We can carry this decoupling to higher order if, instead of following (28), we seek a block diagonal

$$U_{fj} = \begin{pmatrix} U_{fj}^{11} & 0 \\ 0 & U_{fj}^{22} \end{pmatrix}, \quad j > 0 \tag{38}$$

corresponding to an off-diagonal

$$P_j' = \begin{pmatrix} 0 & P_j^{12} \\ P_j^{21} & 0 \end{pmatrix} \tag{39}$$

Then (26) is equivalent to a linear system of the form

$$\left.\begin{aligned} U_{f0}^{11} P_j^{12} - P_j^{12} U_{f0}^{22} &= \beta_{j-1}^{12} \\ U_{f0}^{22} P_j^{21} - P_j^{21} U_{f0}^{11} &= \beta_{j-1}^{21} \\ U_{fj}^{11} &= U_{f0}^{12} P_j^{21} + \beta_{j-1}^{11} \\ U_{fj}^{22} &= - P_j^{21} U_{f0}^{12} + \beta_{j-1}^{22} \end{aligned}\right\} \tag{40}$$

where the $\beta_{j-1}^{ii}$'s are known successively. Since $U_{f0}^{11}$ and $U_{f0}^{22}$ are nearly triangular with no eigenvalues in common, unique determinations for $P_j^{12}$ and $P_j^{21}$ will follow readily by solving the first two systems by forward and back substitution. The last equations then directly specify $U_{fj}$, so that fast growing and fast decaying modes are decoupled to this order. An analogous procedure should apply when the eigenvalues of $U_{f0}$ can be split into stable, purely imaginary, and unstable subsets throughout [0,1] (cf. Hoppensteadt and Miranker (1976) and Sacker and Sell (1980) for related theoretical work).

Summarizing (10), (37), and (38), we can rewrite our transformed system matrix $U(t,\varepsilon)$ in the block form

$$U(t,\varepsilon) = \begin{pmatrix} U^{11} & U^{12} & U^{13} \\ 0 & U^{22} & U^{23} \\ 0 & 0 & \varepsilon U^{33} \end{pmatrix} \tag{41}$$

where all entries are smooth, $U^{12}$ is independent of $\varepsilon$, and $U^{11}$ and $-U^{22}$ have their limiting eigenvalues in the right half plane. Any associated fundamental matrix $Y(t,\varepsilon)$ will satisfy the matrix equation

$$\varepsilon \dot{Y} = U(t,\varepsilon)Y \tag{42}$$

so it is natural to take $Y$ in the block form

$$Y = \begin{pmatrix} Y^{11} & Y^{12} & Y^{13} \\ 0 & Y^{22} & Y^{23} \\ 0 & 0 & Y^{33} \end{pmatrix} \tag{43}$$

To assure the boundedness of $Y$ throughout [0,1], we must choose boundary conditions to reflect the conditional stability of the system. Thus, we'll define the sub-blocks by the initial and terminal value problems

$$\left.\begin{aligned}
\varepsilon\,\dot{Y}^{11} &= U^{11}\,Y^{11},\quad Y^{11}(1) = I_k\\
\varepsilon\,\dot{Y}^{12} &= U^{11}\,Y^{12} + U^{12}\,Y^{22},\quad Y^{12}(1) = 0\\
\varepsilon\,\dot{Y}^{13} &= U^{11}\,Y^{13} + U^{12}\,Y^{23} + U^{13}\,Y^{33},\quad Y^{13}(1) = 0\\
\varepsilon\,\dot{Y}^{22} &= U^{22}\,Y^{22},\quad Y^{22}(0) = I_\ell\\
\varepsilon\,\dot{Y}^{23} &= U^{22}\,Y^{23} + U^{23}\,Y^{33},\quad Y^{23}(0) = 0\\
\dot{Y}^{33} &= U^{33}\,Y^{33},\quad Y^{33}(0) = I_q
\end{aligned}\right\}\tag{44}$$

Asymptotic solutions for these systems appropriate as $\varepsilon \to 0$ are easily obtained using standard singular perturbations (cf. O'Malley (1974)). Indeed, $Y^{11}$ and $Y^{12}$ decay rapidly to zero away from $t = 1$ while $Y^{13}$ tends to its outer limit $-(U^{22})^{-1}(U^{12}Y^{23} + U^{13}Y^{33})$ there. Likewise, $Y^{22}$ decays to zero away from $t = 0$ while $Y^{23}$ tends to $-(U^{22})^{-1}\,U^{23}Y^{33}$ there. $Y^{33}$ follows easily from direct integration since $U^{33}$ is smooth. For moderate values of $\varepsilon$, numerical solutions $Y^{ij}$ can be easily obtained using a good stiff system package or special numerical methods (cf. Söderlind (1981)). Small stepsizes will be required near $t = 0$ ($t = 1$) to resolve the boundary layer behavior of the $Y^{2j}$'s ($Y^{1j}$'s). Approximations in terms of boundary layer corrections follow as in Flaherty and O'Malley (1982).

We shall suppose that a bounded particular solution to the nonhomogeneous system (35) can be obtained, using boundary conditions which reflect the conditional stability of $U_{f0}$ (cf. Mattheij and Staarink (1982)). Then, we can restrict attention to obtaining a complementary solution $y_c$ to the problem (35)-(36) with $f = 0$. A unique solution of the form

$$y_c(t,\varepsilon) = Y(t,\varepsilon)D^{-1}(\varepsilon)c \tag{45}$$

will occur if and only if the matrix

$$D(\varepsilon) = M_0\,T(0,\varepsilon)Y(0,\varepsilon) + M_1\,T(1,\varepsilon)Y(1,\varepsilon) \tag{46}$$

is nonsingular. Indeed, the norm of $D^{-1}c$ as $\varepsilon \to 0$ measures the conditioning of our singularly perturbed boundary value problem and determines the size of its complementary solution. The special structure of $Y(0,\varepsilon)$ and $Y(1,\varepsilon)$, especially as $\varepsilon \to 0$, can be used to asymptotically simplify the invertibility condition. Indeed, the limiting behavior as $\varepsilon \to 0$ within $(0,1)$ can be determined in terms of a reduced boundary value problem consisting of the limiting equation

$$z' = U^{33}(t,0)z \tag{47}$$

and a subset of q of the original boundary conditions (cf. O'Malley (1969), Harris (1973), and Ferguson (1975)). Direct numerical solution of the original boundary value problem (1)-(2) would quite likely break down for moderate $\varepsilon$ values. The special purpose methods of Flaherty and Mathon (1980), Flaherty and O'Malley (1982), Ascher and Weiss (1982), and Weiss (1982) would be more successful.

## 4. The Riccati Method for Special Systems

When we are so fortunate that our system is given in the traditional singular perturbation form

$$\left.\begin{aligned} \varepsilon\dot{x}_1 &= A_{11}(t)x_1 + A_{12}(t)x_2 \\ \dot{x}_2 &= A_{21}(t)x_1 + A_{22}(t)x_2 \end{aligned}\right\} \tag{48}$$

with $A_{11}$ invertible throughout $0 \le t \le 1$, the p vector $x_1$ is already identified as (potentially) fast compared to the q vector $x_2$. Instead of using the previous transformation (33), it is now more convenient to change variables through

$$\begin{pmatrix} x_1 \\ x_2 \end{pmatrix} = Q(t,\varepsilon) \begin{pmatrix} z_1 \\ z_2 \end{pmatrix} \tag{49}$$

where Q has the near identity, lower block-triangular form

$$Q(t,\varepsilon) = \begin{pmatrix} I_p & 0 \\ \varepsilon P(t,\varepsilon) & I_q \end{pmatrix} \tag{50}$$

(cf. Mattheij (1982)). The transformed problem will then take the upper block-triangular form

$$\left.\begin{aligned} \varepsilon \dot{z}_1 &= B_{11}(t,\varepsilon) z_1 + A_{12}(t) z_2 \\ \dot{z}_2 &= B_{22}(t,\varepsilon) z_2 \end{aligned}\right\} \tag{51}$$

provided the matrix P satisfies the singularly perturbed Riccati equation

$$\varepsilon \dot{P} = - P A_{11} + A_{21} - \varepsilon A_{22} P - \varepsilon P A_{12} P \tag{52}$$

Moreover, we will then have

$$B_{11} = A_{11} + \varepsilon A_{12} P \tag{53}$$

$$B_{22} = A_{12} - P A_{22} \tag{54}$$

Here, we shall obtain a smooth power series solution

$$P(t,\varepsilon) = P_0(t) + \varepsilon P_1 (t) + \varepsilon^2 P_2(t) + \dots \tag{55}$$

by iterating in the equation

$$P = A_{11}^{-1} A_{21} - \varepsilon A_{11}^{-1} (A_{22}P + P A_{12} P + \dot{P}). \tag{56}$$

Clearly,

$$P_0 = A_{11}^{-1} A_{21},$$

$$P_1 = A_{11}^{-1}(A_{22}P_0 + P_0A_{12}P_0 + \dot{P}_0),$$

etc. while

$$B_{11} = A_{11} + \varepsilon A_{12}A_{11}^{-1}A_{21} + \varepsilon^2 A_{12}P_1 + \dots$$

and

$$B_{22} = A_{12} - A_{11}^{-1}A_{21}A_{22} - \varepsilon P_1A_{22} + \dots .$$

It is important to observe that the smooth system for $z_2$ will have a (purely) slowly-varying solution and that $z_2(t,0)$ coincides with the outer solution (which is directly obtainable by setting $\varepsilon = 0$ in (48)). The transformation (50) has this substantial advantage compared to the alternative transformation often defined through a matrix of the form

$$\begin{pmatrix} I_p & \varepsilon R(t,\varepsilon) \\ 0 & I_q \end{pmatrix} .$$

It would, instead, completely decouple the fast from the slow modes. Here, $z_1$ involves both fast and slow dynamics. A second transformation

$$z_1 = M(t,\varepsilon)w \tag{57}$$

would be convenient to put the fast system into nearly triangular form. The Schur decomposition of $A_{11}$ could be used to advantage to obtain $A_{11}^{-1}$ and to solve the resulting linear system for M as a power series in $\varepsilon$.

References

1. R. C. Aiken, editor, Proceedings, International Conference on Stiff Computation, Park City, 1982.

2. L. R. Anderson, Decoupling and Reduced Order Modeling of Two-Time-Scale Control Systems, doctoral dissertation, University of Arizona, Tucson, 1979.

3. U. Ascher and R. Weiss, "Collocation for singular perturbation problems II: Linear first order systems without turning points," Technical Report 82-4, Department of Computer Science, University of British Columbia, Vancouver, 1982.

4. J. H. Chow, B. Avramovic, P. V. Kokotovic, and J. R. Winkelman, "Singular perturbations, coherency and aggregation of dynamic systems," Report, Electric Utility Systems Engineering Department, General Electric Company, Schenectady, 1981.

5. W. E. Ferguson, Jr., A Singularly Perturbed Linear Two-Point Boundary Value Problem, doctoral dissertation, California Institute of Technology, Pasadena, 1975.

6. J. E. Flaherty and W. Mathon, "Collocation with polynomial and tension splines for singularly perturbed boundary value problems," SIAM J. Sci. Stat. Comput. 1 (1980), 260-289.

7. J. E. Flaherty and R. E. O'Malley, Jr., "Asymptotic and numerical methods for vector systems of singularly perturbed boundary value problems," Proceedings, Army Numerical Analysis and Computer Conference, Vicksburg, 1982.

8. H. Gingold and P. -F. Hsieh, "On global simplification of a singularly perturbed system of linear ordinary differential equations," preprint, West Virginia and Western Michigan Universities, 1982.

9. G. H. Golub and J. H. Wilkinson, "Ill-conditioned eigensystems and the computation of the Jordan canonical form," SIAM Review 18 (1976), 578-619.

10. W. A. Harris, Jr., "Singularly perturbed boundary value problems revisited," Lecture Notes in Math. 312, Springer-Verlag, Berlin, 1973, 54-64.

11. D. L. Hetrick et al., "Solution methods for simulation of nuclear power systems," Report NP-1928, Electric Power Research Institute, Palo Alto, 1981.

12. F. C. Hoppensteadt and W. L. Miranker, "Differential equations having rapidly changing solutions: Analytic methods for weakly nonlinear systems," J. Differential Equations 22 (1976), 237-249.

13. V. C. Klema and A. J. Laub, "The singular value decomposition: Its computation and some applications," IEEE Trans. Automatic Control 25 (1980), 164-176.

14. B. Kreiss and H. -O. Kreiss, "Numerical methods for singular perturbation problems," SIAM J. Numer. Anal. 18 (1981), 262-276.

15. R. M. M. Mattheij, "Riccati-type transformations and decoupling of singularly perturbed ODE," Proceedings, International Conference on Stiff Computation, Park City, 1982.

16. R. M. M. Mattheij and R. E. O'Malley, Jr., "Decoupling of boundary value problems for two-time-scale systems," Proceedings, International Conference on Stiff Computation, Park City, 1982.

17. R. M. M. Mattheij and G. W. M. Staarink, "An efficient algorithm for solving general linear two point BVP," preprint, Rensselaer Polytechnic Institute and Katholieke Universiteit, Nijmegen, 1982.

18. R. E. O'Malley, Jr., "Boundary value problems for linear systems of ordinary differential equations involving small parameters," J. Math. Mech. 18 (1969), 835-856.

19. R. E. O'Malley, Jr., Introduction to Singular Perturbations, Academic Press, New York, 1974.

20. R. E. O'Malley, Jr., "Singular perturbations and optimal control," Lecture Notes in Math. 680, Springer-Verlag, Berlin, 1978, 170-218.

21. R. E. O'Malley, Jr., "On multiple solutions of singularly perturbed systems in the conditionally stable case," Singular Perturbations and Asymptotics, R. E. Meyer and S. V. Parter, editors, Academic Press, New York, 1980, 87-108.

22. R. E. O'Malley, Jr. and L. R. Anderson, "Time-scale decoupling and order reduction for linear time-varying systems," Optimal Control Appl. and Methods 3 (1982), 133-153.

23. R. E. O'Malley, Jr. and J. E. Flaherty, "Analytical and numerical methods for nonlinear singular singularly-perturbed initial value problems," SIAM J. Appl. Math. 38 (1980), 225-248.

24. B. Porter and R. Crossley, Modal Control, Taylor and Francis, London, 1972.

25. R. J. Sacker and G. R. Sell, "Singular perturbations and conditional stability," J. Math. Anal. Appl. 76 (1980), 406-431.

26. G. Söderlind, "Theoretical and computational aspects of partitioning in the numerical integration of stiff differential systems," Report 8115, Royal Institute of Technology, Stockholm, 1981.

27. A. B. Vasil'eva and V. F. Butuzov, Singularly Perturbed Equations in the Critical Case, Moscow State University, 1978.

28. W. Wasow, Asymptotic Expansions for Ordinary Differential Equations, Wiley, New York, 1965.

29. W. Wasow, "Topics in the theory of linear ordinary differential equations having singularities with respect to a parameter," Institut de recherche mathématique avancée, Strasbourg, 1978.

30. R. Weiss, "An analysis of the box and trapezoidal schemes for linear singularly perturbed boundary value problems," preprint, Technische Universität Wien, 1982.

# COMPOSITE FEEDBACK CONTROL OF NONLINEAR SINGULARLY PERTURBED SYSTEMS

P. V. Kokotovic
Coordinated Science Laboratory
University of Illinois
1101 W. Springfield Avenue
Urbana, Illinois 61801

J. H. Chow
Electric Utility Systems Engineering
General Electric
Schenectady, New York 12345

## ABSTRACT

This note summarizes the concept and properties of the two-time-scale composite control recently developed in a series of papers by the same authors.

## INTRODUCTION

The composite control method[1,2,3] has been developed for the nonlinear optimal control problem in which the system is singularly perturbed by the presence of a small positive scalar $\mu$,

$$\dot{x} = a_1(x) + A_1(x)z + B_1(x)u, \quad x(0) = x_o \tag{1}$$

$$\mu\dot{z} = a_2(x) + A_2(x)z + B_2(x)u, \quad z(0) = z_o \tag{2}$$

where $x \in R^n$, $z \in R^m$, $u \in R^r$ and cost to be optimized is

$$J = \int_0^\infty [p(x)+s'(x)z+z'Q(x)z+u'R(x)u]dt. \tag{3}$$

Assumption I. There exists a domain $D \subset R^n$, containing the origin as an interior point, such that for all $x \in D$ functions $a_1, a_2, A_1, A_2, A_2^{-1}, B_1, B_2$, p, s, and Q are differentiable with respect to x; $a_1, a_2, p$, and s are zero only at x=0; Q and R are positive-definite matrices for all $x \in D$; the scalar $p+s'z+a'Qz$ is a positive-definite function of its arguments x and z, that is, it is positive except for x=0, z=0 where it is zero.

The usual approach to this problem would be to assume that a differentiable optimal value function $V(x,z,\mu)$ exists satisfying Bellman's principle of optimality

$$0 = \min_u [p+s'z+z'Qz+u'Ru+V_x(a_1+A_1z+B_1u)+\frac{1}{\mu}V_z(a_2+A_2z+B_2u)] \tag{4}$$

where $V_x, V_z$ denote the partial derivatives of V. Since the control minimizing (4) is

$$u = -\frac{1}{2}R^{-1}(B_1'V_x'+\frac{1}{\mu}B_2'V_2'), \tag{5}$$

the problem would consist of solving the Hamilton-Jacobi equation

$$\begin{aligned} o = {} & p+s'z+z'Qz+V_x(a_1+A_1z)+\frac{1}{\mu}V_z(a_2+A_2z) \\ & -\frac{1}{4}(V_xB_1+\frac{1}{\mu}V_zB_2)R^{-1}(B_1'V_x'+\frac{1}{\mu}B_2'V_2'), \quad V(0,0,\mu) = 0. \end{aligned} \tag{6}$$

To solve (6) is difficult even for well-behaved nonlinear systems. The presence of $1/\mu$ terms increases the difficulties.

The composite control method avoids these difficulties; in contrast, it takes advantage of the fact that as $\mu \to 0$ the slow and the fast phenomena separate. Instead of dealing with the full problem directly, the method starts by defining two separate lower dimensional subproblems. The solutions of the two subproblems are combined into a composite control whose stabilizing and near optimal properties can be guaranteed.

SLOW AND FAST SUBPROBLEMS

For the slow subproblem, denoted by subscript "s," the fast transient is neglected, that is

$$\dot{x}_s = a_1(x_s) + A_1(x_s)z_s + B_1(x_s)u_s, \qquad x_s(0) = x_o \tag{7}$$

$$0 = a_2(x_s) + A_2(x_s)z_s + B_2(x_s)u_s \tag{8}$$

and, since $A_2^{-1}$ is assumed to exist,

$$z_s(x_s) = -A_2^{-1}(a_2 + B_2u_s) \tag{9}$$

is eliminated from (7) and (8). Then the slow subproblem is to optimally control the "slow subsystem"

$$\dot{x}_s = a_o(x_s) + B_o(x_s)u_s, \qquad x_s(0) = x_o \tag{10}$$

with respect to "slow cost"

$$J_s = \int_0^\infty [p_o(x_s)+2s_o'(x_s)u_s+u_s'R_o(x_s)u_s]dt \tag{11}$$

where

$$\begin{aligned} a_o &= a_1 - A_1A_2^{-1}a_2 \\ B_o &= B_1 - A_1A_2^{-1}B_2 \\ p_o &= p = s'A_2^{-1}a_2 + a_2'A_2'^{-1}QA_2^{-1}a_2 \\ s_o &= B_2'A_2'^{-1}(QA_2^{-1}(QA_2^{-1}a_2 - \tfrac{1}{2}s) \\ R_o &= R + B_2'A_2'^{-1}QA_2^{-1}B_2. \end{aligned} \tag{12}$$

We note that, in view of Assumption I, the equilibrium of the slow subsystem (13) for all $x_s \in D$ is $x_s = 0$, and

$$p_o(x_s) + 2s_o'(x_s)u_s + u_s'R_o(x_s)u_s > 0, \qquad \forall x_s \neq 0, \quad \forall u_s \neq 0. \tag{13}$$

The next assumption concerns the existence of the optimal value function $L(x_s)$ satisfying the optimality principle

$$0 = \min_{u_s}[p_o(x_s)+2s_o'(x_s)u_s + u_s'R_o(x_s)u_s + L_x(a_o(x_s) + B_o(x_s)u_s)] \quad (14)$$

where $L_x$ denotes the derivative of L with respect to its argument $x_s$. The elimination of the minimizing control

$$u_s = - R_o^{-1}(s_o + \frac{1}{2} B_o'L_x') \quad (15)$$

from (14) results in the Hamilton-Jacobi equation

$$0 = (p_o-s_o'R_o^{-1}s_o) + L_x(a_o-B_oR_o^{-1}s_o) - \frac{1}{2} L_xB_oR_o^{-1}B_o'L_x', \qquad L(0) = 0 \quad (16)$$

where $p_o-s_o'R_o^{-1}s_o$ is positive definite in D.

Assumption II. For all $x_s \in D$, (16) has a unique differentiable positive-definite solution $L(x_s)$ with the property that positive constants $k_1,k_2$, $k_3,k_4$ exist such that

$$k_1L_xL_x' \leq - L_x\bar{a}_o \leq k_2L_xL_x' \quad (17)$$

$$k_3\bar{a}_o'\bar{a}_o \leq - L_x\bar{a}_o \leq k_4\bar{a}_o'\bar{a}_o \; . \quad (18)$$

Then $L(x_s)$ is a Lyapunov function guaranteeing the asymptotic stability of $x_s = 0$ for the slow subsystem (10) controlled by (15), that is, for the feedback system

$$\dot{x}_s = a_o - B_oR_o^{-1}(s_o + \frac{1}{2} B_o'L_x') = \bar{a}_o(x_s). \quad (19)$$

It also guarantees that D belongs to the region of attraction of $x_s = 0$.

For the fast subproblem, denoted by subscript "f," we recall that only an $O(\mu)$ error is made by replacing x with $x_s$, or z with $z_s$. Thus, we subtract (8) from (2), introduce $z_f = z-z_s$, $u_f=u-u_s$, neglect $O(\mu)$ terms, and define the fast subproblem as

$$\mu\dot{z}_f = A_2(z)z_f + B_2(x)u_f, \qquad z_f(0) = z_o-z_s(0), \quad (20)$$

$$J_f = \int_0^\infty (z_f'Q(x)z_f + u_f'R(x)u_f)dt. \quad (21)$$

This problem is to be solved for every fixed $x \in D$. It has the familiar linear quadratic form and a controllability assumption is natural.

Assumption III. For every fixed $x \in D$,

$$\text{rank } [B_2, A_2B_2, \ldots, A_2^{m-1}B_2] = m. \tag{22}$$

Alternatively, a less demanding stabilizability assumption can be made. For each $x \in D$ the optimal solution of the fast subproblem is

$$u_f(z_f, x) = -R^{-1}(x)B_2'(x)K_f(x)z_f \tag{23}$$

where $K_f(x)$ is the positive-definite solution of the x-dependent Riccati equation

$$0 = K_fA_2 + A_2'K_f - K_fB_2R^{-1}B_2'K_f + Q. \tag{24}$$

The control (23) is stabilizing in the sense that the fast feedback system

$$\mu\dot{z}_f = (A_2 - B_2R^{-1}B_2'K)z_f \triangleq \bar{A}_2(x)z_f \tag{25}$$

has the property that $\text{Re}\lambda[\bar{A}_2(x)] < 0$, $\forall x \in D$.

## COMPOSITE CONTROL

We now form a "composite" control $u_c = u_s + u_f$, in which $x_s$ is replaced by $x$ and $z_f$ by $z + A_2^{-1}(a_2 + B_2u_s(x))$, that is

$$\begin{aligned} u_c(x,z) &= u_s(x) - R^{-1}B_2'K_f(z + A_2^{-1}(a_2 - B_2u_s(x)) \\ &= -R_o^{-1}(s_o + \tfrac{1}{2}B_o'L_x') - R^{-1}B_2'K_f(z + \bar{A}_2^{-1}\bar{a}_2) \end{aligned} \tag{26}$$

where

$$\bar{a}_2(x) = a_2 - \tfrac{1}{2}B_2R^{-1}(b_1'L_x' + B_2'V_1), \qquad \bar{a}_2(0) = 0$$

$$V_1' = -(s' + 2a_2'K_f + L_x\bar{A}_1)\bar{A}_2^{-1} \tag{27}$$

$$\bar{A}_1 = A_1 - B_1R^{-1}B_2'K_f.$$

The properties of the system controlled by the composite control are summarized in the following theorem.

Theorem. When Assumptions I, II, and III are satisfied then there exists $\mu^*$ such that $\forall\ \mu \in (0,\mu^*]$, the composite control $u_c$ defined by (26) stabilizes the full system (1), (2) in a sphere centered at x=0, z=0. The corresponding cost $J_c$ is bounded. Moreover, $J_c$ is near optimal in the sense that $J_c \to J_s$ as $\mu \to 0$.

This theorem shows that the considered nonlinear regulator problem is well-posed with respect to $\mu$. It is the basis for a two-time scale design procedure whose steps are illustrated by the following example.

Example. The system and the cost are

$$\dot{x} = -\frac{3}{4}x^3 + z \tag{28}$$

$$\mu\dot{z} = -z + u \tag{29}$$

$$J = \int_0^\infty (x^6 + \frac{3}{4}z^2 + \frac{1}{4}u^2)dt. \tag{30}$$

Step 1. The slow subproblem

$$\dot{x}_s = -\frac{3}{4}x_s^3 + u_s \tag{31}$$

$$J_s = \int_0^\infty (x_s^6 + u_s^2)dt \tag{32}$$

consists in solving the Hamilton-Jacobi equation

$$L_x = \frac{dL}{dx_s} = x_s^3, \qquad L(0) = 0 \tag{33}$$

which yields

$$L = \frac{1}{4}x_s^4, \quad u_s = -\frac{1}{2}x_s^3, \quad \dot{x}_s = -\frac{5}{4}x_s^3. \tag{34}$$

Step 2. Assuumption II

$$k_1x_s^6 \le \frac{5}{4}x_s^6 \le k_2x_s^6, \tag{35}$$

$$\frac{25}{16}k_3x_s^6 \le \frac{5}{4}x_s^6 \le \frac{25}{16}k_4x_s^6, \tag{36}$$

is satisfied by

$$k_1 = k_2 = \frac{5}{4}, \qquad k_3 = k_4 = \frac{4}{5}. \tag{37}$$

Step 3. The fast subproblem

$$\mu\dot{z}_f = - z_f + u_f \tag{38}$$

$$J_f = \int_0^\infty (\frac{3}{4} z_f^2 + \frac{1}{4} u_f^2)dt \tag{39}$$

is, in this case, independent of x and its solution is

$$K_f = \frac{1}{4}, \qquad u_f = - z_f, \qquad \mu\dot{z}_f = - 2z_f. \tag{40}$$

Step 4. The design is completed by forming the composite control

$$u_c = - x^3 - z \tag{41}$$

and applying it to the full system (28), (29), that is

$$\dot{x} = - \frac{3}{4} x^3 + z \tag{42}$$

$$\mu\dot{z} = - x^3 - 2z.$$

It should be noted that this system could not have been designed by methods based on linearization since its linearized model at $x = 0$, $z = 0$ has a zero eigenvalue. However, the composite control theorem guarantees that the equilibrium $x = 0$, $z = 0$ is asymptotically stable for $\mu$ sufficiently small. A bound for $\mu$, estimates of the stability region, and further details are contained in [1,2,3].

REFERENCES

1. Chow, J. H. and P. V. Kokotovic, A decomposition of near-optimum regulators for systems with slow and fast modes, IEEE Trans. on Auto. Cont., AC-21, 5, 701, 1976.
2. Chow, J. H. and P. V. Kokotovic, Near-optimal feedback stabilization of a class of nonlinear singularly perturbed systems, SIAM J. Cont. and Opt., 16, 756, 1978.
3. Chow, J. H. and P. V. Kokotovic, A two-stage Lyapunov-Bellman feedback design of a class of nonlinear systems, IEEE Trans. on Auto. Cont., AC-26, 656, 1981.

This work was supported in part by the National Science Foundation under Grant ECS-79-19396, in part by the Joint Services Electronics Program under Contract N00014-79-C-0424, and in part by the U.S. Air Force under Grant AFOSR 78-3633.

# REGULAR PERTURBATIONS IN OPTIMAL CONTROL

A. BENSOUSSAN (*)

## INTRODUCTION.

We study in this paper a problem of regular perturbations in Optimal control. This problem has already been considered by J. Cruz [1]. However the treatment of Cruz is limited to the case without constraints on the control. The main objective of this paper is to extend the theory to get rid of this restriction. Our method of proof is also different.

( ) University Paris-Dauphine and INRIA.

This work has been partly supported by DOE Office of Electric Energy Systems under contract 01-80, RA-50 154.

## 1 - SETTING OF THE PROBLEM.

### 1.1 - Notation - Assumptions.

We consider functions $g_0$, $g_1$ such that

$$(1.1)\quad \begin{aligned} & g_0(x,v),\ g_1(x,v) : R^n \times R^k \to R^n \\ & |g_i(x,v)| \le \bar{g}_i(1 + |x| + |v|) \end{aligned}$$

$g_0$ is $C^3$ in $x,v$ with Lipschitz third derivatives

$g_1$ is $C^2$ in $x,v$ with Lipschitz second derivatives.

The derivatives of $g_0$, $g_1$ are bounded.

Let also $f_0$, $f_1$, $h_0$, $h_1$ be such that

$$(1.2)\quad \begin{aligned} & f_0(x,v),\ f_1(x,v) : R^n \times R^k \to R \\ & h_0(x),\ h_1(x) : R^n \to R \end{aligned}$$

$f_0$, $h_0$ are $C^3$ functions with Lipschitz third derivatives,

$f_1$, $h_1$ are $C^2$ functions with Lipschitz second derivatives

$$|f_i(x,v)| \le \bar{f}_i(1 + |x|^2 + |v|^2)$$

$$|h_i(x)| \le \bar{h}_i(1 + |x|^2).$$

We consider the following control problem

$$(1.3)\quad \begin{aligned} & \dot{x} = g_0(x,v) + \varepsilon g_1(x,v) \\ & x(o) = x \end{aligned}$$

An admissible control will be a function $v(.)$ in $L^2(o,T\ ;\ R^k)$ such that $v(t) \in U_{ad}$ a.e. where

(1.4) $\quad U_{ad}$ closed convex subset of $R^k$ (non empty).

We minimize the cost function

$$(1.5) \qquad J_\varepsilon(v(.)) = \int_0^T [f_0(x(t),v(t)) + \varepsilon f_1(x(t),v(t))]\, dt$$

$$+ h_0(x(T)) + \varepsilon h_1(x(T)).$$

Our objective is to study the behaviour of inf $J_\varepsilon(v(.))$ as $\varepsilon$ tends to 0.

1.2 - Necessary conditions.

Although we shall not require explicitly the existence of an optimal control for (1.5), it is useful to state the necessary conditions of optimality. Denoting by $u_\varepsilon$ an optimal control and by $y_\varepsilon$ the corresponding trajectory, we obtain

$$(1.6) \qquad \dot{y}_\varepsilon = g_0(y_\varepsilon,\ u_\varepsilon) + \varepsilon g_1(y_\varepsilon,\ u_\varepsilon) \qquad\qquad \dot{y}_\varepsilon(o) = x$$

$$(1.7) \qquad -\dot{p}_\varepsilon = (g_{0x}(y_\varepsilon,\ u_\varepsilon) + \varepsilon g_{1x}(y_\varepsilon,\ u_\varepsilon))p_\varepsilon + f_{0x}(y_\varepsilon,\ u_\varepsilon) + \varepsilon f_{1x}(y_\varepsilon,u_\varepsilon)$$

$$p_\varepsilon(T) = h_{0x}(y_\varepsilon\ (T)) + \varepsilon h_{1x}(y_\varepsilon,(T))$$

$$(1.8) \qquad (H_{0v}(y_\varepsilon(t),\ p_\varepsilon(t),\ u_\varepsilon(t)) + \varepsilon\ H_{1v}(y_\varepsilon(t),\ p_\varepsilon(t),\ u_\varepsilon(t))).(v - u_\varepsilon(t)) \geq 0$$

$$\text{a.e.t}\ , \quad \forall v \in U_{ad}.$$

For $\varepsilon = o$, we get

$$(1.9) \qquad \dot{y}_0 = g_0(y_0,u_0) \qquad\qquad y_0(o) = x$$

$$(1.10) \qquad -\dot{p}_0 = g^*_{0x}(g_0,u_0)\, p_0 + f_{0x}(y_0,u_0)$$

$$p_0(T) = h_{0x}(y_0(T))$$

$$(1.11) \qquad H_{0v}(y_0(t), p_0(t), u_0(t)) \,.\, (v - u_0(t)) \geq o$$

$$\text{a.e.}\, t \;, \quad \forall v \in U_{ad} \,.$$

In (1.8), (1.11) $H_0$, $H_1$ represent the Hamiltonians.

$$(1.12) \qquad H_0(x,p,v) = f_0(x,v) + p.\, g_0(x,v)$$

$$H_1(x,p,v) = f_1(x,v) + p.\, g_1(x,v).$$

We shall assume that there exists a triplet $(y_0,u_0,t_0)$ satisfying (1.9), (1.10), (1.11) such that in addition

$$(1.13) \qquad H_{0vv}(x,p_0(t),v) \geq \beta_0 I, \; \forall t \in [o,T] \;, \; \forall v, \; \forall x$$

in a bounded set

$$(1.14) \qquad H_{0xv} - H_{0xv}\, H^{-1}_{ovv}\, H_{vx} \geq \beta_0 I$$

$$(1.15) \qquad h_{0xx}(x) \geq o$$

where in (1.14), (1.15) the conditions on the arguments are the same as in (1.13).

We note the

Theorem 1.1 : We assume (1.9), (1.10), (1.11) and (1.13), (1.14), (1.15) then $u_o$ is an optimal control for $J_o(v(.))$ and the unique optimal control.

Proof :

Let v be any control and z be the solution of

$$\dot{z} = g_o(z,v) \qquad z(o) = x$$

Set $\tilde{u} = v - u_o \,, \quad \tilde{y} = z - y_o \,.$

We have

$$J_o(v(.)) = \int_0^T f_o(y_o + \tilde{y},\ u_o + \tilde{u})dt + h_o(y_o(T) + \tilde{y}(T))$$

$$= J_o(u_o) + \int_0^T (f_{ox}(y_o,u_o)\tilde{y} + f_{ov}(y_o,u_o)\tilde{u})dt$$

(1.16)
$$+ h_{ox}(y_o(T))\ \tilde{y}(T)$$

$$+ \int_0^T \int_0^1 \int_0^1 \lambda\ [f_{oxx}(y_o + \mu\lambda\tilde{y},\ u_o + \mu\lambda\tilde{u})\tilde{y}.\tilde{y} + f_{ovv}(\ )\tilde{u}.\tilde{u}$$

$$+ 2\ f_{oxv}(\ )\tilde{u}\ \tilde{y}]\ d\mu d\lambda dt + \int_0^1 \int_0^1 \lambda h_{oxx}(y_o(T) + \mu\lambda\tilde{y}(T))\ \tilde{y}(T).\tilde{y}(T)$$

Moreover from (1.10), (1.11)

$$\int_0^T [f_{ox}(y_0,u_0)\tilde{y} + f_{ov}(y_0,u_0)\tilde{u}]\, dt + h_{ox}(y_0(T))\, \tilde{y}\, (T)$$

$$\geq \int_0^T [(-\dot{p}_0 - g_{ox}^* \, p_0)\tilde{y} - g_{ov}^* \, p_0.\tilde{u}]\, dt + p_0(T)\tilde{y}(T)$$

$$= \int_0^T p_0.(g_0(y_0 + \tilde{y},\, u_0 + \tilde{u}) - g_0(y_0,u_0) - g_{ox}\,\tilde{y} - g_{ov}\,\tilde{u})\, dt$$

which combined with (1.16) yields

$$J_0(v) \geq J_0(u_0) + \int_0^T \int_0^1 \int_0^1 \lambda\, [H_{oxx}(y_0 + \mu\lambda\tilde{y},\, u_0 + \mu\lambda\tilde{u})\tilde{y}.\tilde{y}$$

$$+ 2H_{oxv}(\;)\tilde{u}.\tilde{y} + H_{ovv}(\;)\tilde{u}.\tilde{u}]\, d\mu d\lambda dt \geq J_0(u_0)$$

Hence $u_0$ is optimal. Let us prove uniqueness. If $J_0(v) = J_0(u_0)$ then it follows that

$$\text{(1.17)} \qquad \tilde{u} + H_{ovv}^{-1}\, H_{ovx}\tilde{y} = 0.$$

But

$$\frac{d\tilde{y}}{dt} = g_0(y_0 + \tilde{y},\, u_0 + \tilde{u}) - g_0(y_0,u_0)$$

$$\tilde{y}(o) = 0$$

which with (1.17) using Gronwall's inequality implies $\tilde{y} = 0$, $\tilde{u} = 0$ and the desired result is proved.

## 2 - ASYMPTOTICS.

### 2.1 - Formal expansion.

We approximate $u_\varepsilon$ by $u_0 + \varepsilon u_{1\varepsilon}$ , where $u_{1\varepsilon}$ is such that

$u_0 + \varepsilon u_{1\varepsilon} \in U_{ad}$, and $y_\varepsilon$ by $y_0 + \varepsilon y_{1\varepsilon}$, $p_\varepsilon$ by $p_0 + \varepsilon p_{1\varepsilon}$. From (1.6), (1.7), (1.8) we get up to terms of order $\varepsilon^2$

(2.1) $$\dot{y}_{1\varepsilon} = g_{ox}(y_0,u_0)y_{1\varepsilon} + g_{ov}(y_0,u_0)u_{1\varepsilon} + g_1(y_0,u_0)$$

$$y_{1\varepsilon}(0) = 0$$

(2.2) $$-\dot{p}_{1\varepsilon} = g^*_{ox}\, p_{1\varepsilon} + H_{oxx}\, y_{1\varepsilon} + H_{oxv}\, u_{1\varepsilon} + H_1$$

$$p_{1\varepsilon}(T) = h_{oxx}\, y_{1\varepsilon}(T) + h_{1x}$$

(2.3) $$[\varepsilon(H_{ovx}\, y_{1\varepsilon} + g^*_{ov}\, p_{1\varepsilon} + H_{ovv}\, u_{1\varepsilon} + H_{1v}) + H_{ov}].(w - (u_0 + \varepsilon u_{1\varepsilon}))\geq 0$$

$$\forall w \in U_{ad}\ ,\ \text{a.e.}$$

In fact (2.1), (2.2), (2.3) express necessary conditions of optimality for the following optimal control problem

(2.4) $$\min_{v(.)} \varepsilon J_1(v) + 2\int_0^T H_{ov}(y_0,p_0,u_0)\, v\, dt,$$

$$u_0(t) + \varepsilon v(t) \in U_{ad}\ ,\quad \text{a.e.}$$

where $J_1(v)$ is defined by

$$J_1(v) = \frac{1}{2}\{\int_0^T (H_{oxx}(y_0,p_0,u_0)z.z + 2\, H_{oxv}\, v.z$$

(2.5) $$+ H_{ovv}\, v.v + 2\, H_{1v}\, v + 2\, H_{1x} z)dt$$

$$+ h_{oxx}(y_0(T))\, z\,(T).z(T) + 2h_{1x}(y_0(T)).z(T)\}\ ,$$

and

(2.6) $$\dot{z} = g_{ox}(y_o,u_o)z + g_{ov}(y_o,u_o)v + g_1(y_o,u_o)$$

$$z(o) = 0$$

We obtain a linear quadratic control problem with constraints. When $U_{ad}$ is the whole space, we notice that

$$H_{ov}(y_o,p_o,u_o) = 0$$

and then (2.4) amounts to $\text{Min}\ J_1(v)$.

Reasoning like in Theorem 1.1, one checks that the problem (2.5), (2.6) has one and only one solution $u_{1\varepsilon}$.

2.2 -Statement of the main results.

Let us define the functional

$$L_1(v) = \frac{1}{6}\int_0^T [H_{oxxx}(y_o,p_o,u_o)z.z.z + 3\,H_{oxxv}vzz$$

$$+ 3\,H_{oxvv}vvz + H_{ovvv}vvv + 3\,H_{opxx}zzq + 6\,H_{opxv}vzq + 3\,H_{opvv}vvq]dt$$

$$+ \frac{1}{2}\int_0^T [H_{1xx}(y_o,p_o,u_o)zz + 2\,H_{1xv}vz + H_{1vv}vv$$

$$+ 2(H_{1pv}vq + H_{1px}zq)]dt + \frac{1}{6}\,h_{oxxx}(y_o(T))z(T).z(T).z(T) +$$

$$+ \frac{1}{2}\,h_{1xx}(u_o(T))z(T).z(T)$$

where z has been defined by (2.6) and q is defined by

$$- \dot{q} = g_{0x}^{*}(y_0,u_0)q + H_{0xx}(y_0,p_0,u_0)z + H_{0xv}(y_0,p_0,u_0)v +$$

$$(2.8) \qquad + H_{1x}(y_0,p_0,u_0)$$

$$q(T) = h_{0xx}(y_0(T))z(T) + h_{1x}(y_0(T))$$

Our main result is the following.

Theorem 2.1 : We assume (1.1), (1.2), (1.4) and (1.13), (1.14), (1.15). Then we have the estimates

$$(2.9) \qquad \begin{aligned} &|\inf_{\varepsilon} J_\varepsilon(v(.)) - J_0(u_0) - \varepsilon[\int_0^T H_1(y_0,p_0,u_0)dt + H_1(y_0(T)) \\ &+ \int_0^T H_{0v}u_{1\varepsilon}dt] - \varepsilon^2 J_1(u_{1\varepsilon}) - \varepsilon^3 L_1(u_{1\varepsilon})| \leq C\varepsilon^4 \end{aligned}$$

$$(2.10) \qquad |\inf_{\varepsilon} J_\varepsilon(v(.)) - J_\varepsilon(u_0 + \varepsilon u_{1\varepsilon})| \leq C\varepsilon^4 .$$

## 3 - PROOF OF THE MAIN RESULT.

### 3.1 - Preliminary results.

We consider $u_\varepsilon$ such that

$$(3.1) \qquad J_\varepsilon(u_\varepsilon) \leq J_\varepsilon (u_0 + \varepsilon u_{1\varepsilon})$$

and we write

$$u_\varepsilon = u_0 + \varepsilon u_{1\varepsilon} + \tilde{u}_\varepsilon$$

$$y_\varepsilon = y_0 + \varepsilon y_{1\varepsilon} + \tilde{y}_\varepsilon \quad .$$

We note that $u_0$ is a bounded function. This follows from (1.11) which yields, using (1.13).

(3.2) $\beta_0|v - u_0(t)|^2 \leq H_{ov}(y_0(t),p_0(t),v).(v - u_0(t)) \quad \forall v \text{ in } U_{ad}.$

Moreover we have

(3.3) $u_{1\varepsilon}, y_{1\varepsilon}, p_{1\varepsilon}$ are bounded in $L^\infty$, independantly of $\varepsilon$.

Indeed, we first note that

$$H_{ov}(y_0,p_0,u_0)v \geq 0 \qquad \forall\, v(.) \text{ such that}$$

$$u_0(t) + \varepsilon v(t) \in U_{ad} \quad \text{a.e,}$$

and 0 is admissible. Therefore

(3.4) $J_1(u_{1\varepsilon}) \leq J_1(o)$

From (2.5) and (2.6), and the assumptions (1.13), (1.14), (1.15), one can then deduce that $u_{1\varepsilon}$ remains bounded in $L^2$. This implies from (2.1) and (2.2) that $y_{1\varepsilon}$, $p_{1\varepsilon}$ remain bounded in $L^\infty$. Taking $w = u_0$ in (2.3), we easily deduce

$$(H_{ovx}\, y_{1\varepsilon} + g^*_{ov}\, p_{1\varepsilon} + H_{ovv}\, u_{1\varepsilon} + H_{1v}).u_{1\varepsilon} \leq 0$$

hence

$$|u_{1\varepsilon}(t)| \leq C(|y_{1\varepsilon}(t)| + |p_{1\varepsilon}(t)|).$$

Therefore $u_{1\varepsilon}$ is bounded.

Lemma 3.1. We have

(3.5) $$|\tilde{u}_\varepsilon|_{L^2} \le C\varepsilon^2$$

(3.6) $$|\tilde{y}_\varepsilon|_{C(0,T)} \le C\varepsilon^2$$

Proof

We first have the formulas

(3.7)
$$\begin{aligned}
J_\varepsilon(u_\varepsilon) &= J_0(u_0) + \varepsilon\Big[\int_0^T H_1(y_0,p_0,u_0)dt + h_1(y_0(T))\Big] + \\
&+ \int_0^T H_{0v}(\varepsilon u_1 + \tilde{u}_\varepsilon)dt + \\
&+ \int_0^T\int_0^1\int_0^1 \lambda\,[H_{0xx}(y_0 + \mu\lambda(\varepsilon y_1 + \tilde{y}_\varepsilon),p_0,u_0 + \mu\lambda(\varepsilon u_1 + \tilde{u}_\varepsilon))(\varepsilon y_1 + \tilde{y}_\varepsilon)^2 + \\
&+ H_{0vv}(\varepsilon u_1 + \tilde{u}_\varepsilon)^2 + 2\,H_{0xv}(\varepsilon u_1 + \tilde{u}_\varepsilon)(\varepsilon y_1 + \tilde{y}_\varepsilon)]\,d\mu d\lambda dt \\
&+ \varepsilon\int_0^T\int_0^1 [H_{1x}(y_0 + \lambda(\varepsilon y_1 + \tilde{y}_\varepsilon),p_0,u_0 + \lambda(\varepsilon u_1 + \tilde{u}_\varepsilon))(\varepsilon y_1 + \tilde{y}_\varepsilon) + \\
&+ H_{1v}(\quad)(\varepsilon u_1 + \tilde{u}_\varepsilon)]\,d\lambda dt \\
&+ \frac{1}{2}\int_0^1\int_0^1 h_{0xx}(y_0(T) + \mu\lambda(\varepsilon y_1(T) + \tilde{y}_\varepsilon(T)))(\varepsilon y_1(T) + \tilde{y}_\varepsilon(T))^2 \\
&+ \varepsilon\int_0^1 h_{1x}(y_0(T) + \lambda(\varepsilon y_1(T) + \tilde{y}_\varepsilon(T)))(\varepsilon y_1(T) + \tilde{y}_\varepsilon(T))
\end{aligned}$$

$$(3.8)\qquad \frac{d\tilde{y}_\varepsilon}{dt} = g_0\,(y_0 + \varepsilon y_{1\varepsilon} + \tilde{y}_\varepsilon,\ u_0 + \varepsilon u_{1\varepsilon} + \tilde{u}_\varepsilon) - g_0(y_0,u_0)$$

$$- \varepsilon(g_{0x}\, y_{1\varepsilon} + g_{0v}\, u_{1\varepsilon}) + \varepsilon(g_1(y_0 + \varepsilon y_{1\varepsilon} + \tilde{y}_\varepsilon, u_0 + u_{1\varepsilon} + \tilde{u}_\varepsilon) -$$

$$- g_1(y_0,u_0))$$

$$\tilde{y}_\varepsilon(0) = 0$$

Expanding (3.7) we first use

$$(3.9)\qquad \int_0^T \int_0^1 \int_0^1 \lambda\ [H_{oxx}\ \tilde{y}_\varepsilon^{\ 2} + H_{ovv}\ \tilde{\omega}_\varepsilon^{\ 2} + 2\ H_{oxv}\ \tilde{u}_\varepsilon\ \tilde{y}_\varepsilon] d\mu d\lambda dt$$

$$\geq \beta_0 \int_0^T \int_0^1 \int_0^1 \lambda |\tilde{u}_\varepsilon + H_{vv}^{-1}\ H_{vx}\ \tilde{y}\varepsilon|^2\ d\mu d\lambda dt + \frac{\beta_0}{2} \int_0^T |\tilde{y}_\varepsilon|^2\ dt$$

Let us set

$$(3.10)\qquad \tilde{\psi}_\varepsilon(\mu,\lambda,t) = \tilde{u}_\varepsilon(t) + H_{vv}^{-1}\ H_{vx}(y_0 + \mu\lambda(\varepsilon y_{1\varepsilon} + \tilde{y}_\varepsilon), u_0 + \mu\lambda(\varepsilon u_{1\varepsilon} + \tilde{u}_\varepsilon))\tilde{y}_\varepsilon(t$$

Replacing $\tilde{u}_\varepsilon(t)$ in (3.8), by its value from (3.10) and using Gronwall's inequality we can deduce from (3.8) that

$$(3.11)\qquad |\tilde{y}_\varepsilon(t)|^2 \leq C\ (\int_0^T |\tilde{\psi}_\varepsilon(\mu,\lambda,\ t)|^2 dt + \varepsilon^4)$$

where the constant does not depend on $\mu,\lambda,\varepsilon$.

We next consider estimates of the form

(3.12) $$|\int_0^T\int_0^1\int_0^1 \lambda\, [H_{oxx}(y_0 + \mu\lambda(\varepsilon y_{1\varepsilon} + \tilde{y}_\varepsilon), u_0 + \mu\lambda(\varepsilon u_{1\varepsilon} + \tilde{u}_\varepsilon)) -$$

$$- H_{oxx}(y_0, p_0, u_0)]\varepsilon y_{1\varepsilon} \cdot \tilde{y}_\varepsilon|$$

$$\leq C\varepsilon \int_0^T (|\tilde{y}_\varepsilon(t)|^2 + |\tilde{u}_\varepsilon(t)|^2)dt + C\varepsilon^2 (\int_0^T (|\tilde{y}_\varepsilon(t)|^2 + |\tilde{u}_\varepsilon(t)|^2)dt)^{1/2}$$

$$\leq C\varepsilon \int_0^T |\tilde{y}_\varepsilon(t)|^2 dt + C\varepsilon \int_0^T\int_0^1\int_0^1 \lambda|\tilde{\psi}\varepsilon(\mu,\lambda,t)|^2 d\mu d\lambda dt$$

$$+ C\varepsilon^2 [\int_0^T |\tilde{y}_\varepsilon(t)|^2 dt + \int_0^T\int_0^1\int_0^1 \lambda|\tilde{\psi}\varepsilon|^2 d\mu d\lambda dt]^{1/2} .$$

We also notice that from (2.3)

$$\varepsilon\int_0^T [(H_{oxx}(y_0, p_0, u_0) y_{1\varepsilon} + H_{oxv}\, u_{1\varepsilon} + H_{1x})\, \tilde{y}_\varepsilon$$

$$+ (H_{ovx}\, y_{1\varepsilon} + H_{ovv}\, u_{1\varepsilon} + H_{1v})\tilde{u}_\varepsilon]dt$$

$$+ \varepsilon(h_{oxx}(y_0(T))y_1(T) + h_{1x}(y_0(T)))\, \tilde{y}_\varepsilon(T)$$

$$\geq \varepsilon\int_0^T (-\dot{p}_{1\varepsilon} - g_{ox}\, p_{1\varepsilon})\tilde{y}_\varepsilon dt - \int_0^T H_{ov}\, \tilde{u}_\varepsilon dt - \varepsilon\int_0^T p_{1\varepsilon}\, g_{ov}\, \tilde{u}_\varepsilon dt$$

$$+ \varepsilon p_{1\varepsilon}(T)\, \tilde{y}_\varepsilon(T)$$

$$\geq -\int_0^T H_{ov}\, \tilde{u}_\varepsilon dt + \varepsilon\int_0^T p_{1\varepsilon} \cdot (g_0(y_\varepsilon, u_\varepsilon) - g_0(y_0, u_0)$$

$$- g_{ox}(\varepsilon y_1 + \tilde{y}_\varepsilon) - g_{o\tilde{v}}(\varepsilon u_1 + \tilde{u}_\varepsilon) + \varepsilon(g_1(y_\varepsilon, u_\varepsilon) - g_1(y_0, u_0)))dt$$

Expanding and treating some terms like (3.12), we obtain after collecting terms

$$(3.13)\quad J_\varepsilon(u_\varepsilon) \geq J_o(u_o) + \varepsilon[\int_0^T H_1(y_0,p_0,u_0)dt + h_1(y_0(T))$$

$$+ \int_0^T H_{ov}\, u_{1\varepsilon}\, dt] + \varepsilon^2\, J_1(u_{1\varepsilon}) + \varepsilon^3\, L_1(u_{1\varepsilon})$$

$$+ \frac{\beta 0}{2}\, [\int_0^T \int_0^1 \int_0^1 \lambda|\tilde{\psi}\varepsilon|^2 dt + \int_0^T |\tilde{y}_\varepsilon|^2 dt] - C\varepsilon^4 \;.$$

One then uses (3.1) and express the fact that by a similar calculation

$$(3.14)\quad J_\varepsilon(u_0 + \varepsilon u_{1\varepsilon}) \leq J_o(u_o) + \varepsilon[\int_0^T H_1\, (y_0,p_0,u_0)dt + h_1(y_0(T))$$

$$+ \int_0^T H_{ov}\, u_{1\varepsilon}\, dt] + \varepsilon^2\, J_1(u_{1\varepsilon}) + \varepsilon^3\, L_1(u_{1\varepsilon}) + C\varepsilon^4$$

Compared to (3.13) we get

$$\int_0^T \int_0^1 \int_0^1 \lambda|\tilde{\psi}\varepsilon|^2 dt\; d\lambda d\mu + \int_0^T |\tilde{y}_\varepsilon|^2 dt \leq C\varepsilon^4$$

hence also

$$\int_0^T |\tilde{u}_\varepsilon|^2 dt \leq C\varepsilon^4 \;.$$

Using (3.8) we deduce (3.6).

□

3.2 - Proof of Theorem 2.1.

From (3.13) and (3.14) we deduce that for any $u_\varepsilon$ satisfying (3.1)

$$|J_\varepsilon(u_\varepsilon) - J_o(u_o) - \varepsilon\, [\int_0^T H_1(y_0,p_0,u_0)dt + h_1(y_0(T)) + \int_0^T H_{ov} u_{1\varepsilon} dt]$$

$$- \varepsilon^2\, J_1(u_{1\varepsilon}) - \varepsilon^3\, L_1(u_{1\varepsilon})]| \leq C\varepsilon^4 \;.$$

We can find $u_\varepsilon$ such that

$$J_\varepsilon(u_\varepsilon) \leq \inf J_\varepsilon(v) + C\varepsilon^4$$

and $u_\varepsilon$ satisfies (3.1). From this and (3.14) it is easy to deduce the desired result.

□

Remark 3.1. For a similar treatment for stochastic systems, see A. Bensoussan [2].

REFERENCES.

[1] J. CRUZ , Feedback Systems, Mac Graw Hill, 1972.

[2] A. BENSOUSSAN "On Perturbation Methods in Stochastic Control" presented at $2^{nd}$ Bad Honnef Workshop on Stochastic Differential Systems - University of Bonn - June 28, July 2 1982.

# HIGH GAIN FEEDBACK CONTROL SYNTHESIS

JAN C. WILLEMS

Mathematics Institute
University of Groningen
P.O. Box 800
9700 AV Groningen
The Netherlands

This talk was devoted to an exposition of the theory of 'almost invariant subspaces' which was developed in [1,2]. This theory provides a geometric approach to the synthesis of high gain feedback control synthesis and is therefore intimately related to singular perturbations.

The basic notions in this area, which pertain to the finite dimensional linear time-invariant system

$$\Sigma: \dot{x} = Ax + Bu \; ; \; y = Cx$$

with $x \in X := \mathbb{R}^n$, $u \in U := \mathbb{R}^m$, and $y \in Y := \mathbb{R}^p$, are given in the following definition:

Definition: A subspace $V_a \subset X$ is said to be *almost controlled invariant* if it has the following property: $\forall x_0 \in V_a$ and $\forall \varepsilon > 0$, $\exists \underline{u}: \mathbb{R}^+ \to U$, $\underline{u} \in L_1^{loc}$, such that the solution to $\dot{\underline{x}} = A\underline{x} + B\underline{u}$, $\underline{x}(0) = x_0$ satisfies

$$\sup_{t \geq 0} \inf_{v \in V_a} \| \underline{x}(t) - v \| \leq \varepsilon .$$

A subspace $S_a \subset X$ is said to be *almost conditionally invariant* if it has the following property: $\forall x_0 \in S_a$ and $\forall \varepsilon > 0$, $\exists$ matrices K,L such that the observer $\dot{\underline{w}} = K\underline{w} + L\underline{y}$, $\underline{w}(0) = 0$, for $\dot{\underline{x}} = A\underline{x}$ ; $\underline{y} = C\underline{x}$, $\underline{x}(0) = x_0$, yields $\sup_{t \geq 0} \| \underline{w}(t) - \underline{x}(t) (\text{mod } S_a) \| \leq \varepsilon$.

These are natural generalizations of the notions of controlled invariant ((A,B) - invariant) and conditionally invariant ((A,C) - invariant) subspaces which are the key concepts in the very successful geometric approach in linear system theory (see [3] for a modern exposition). In fact if $\underline{V}, \underline{V}_a, \underline{S}$, and $\underline{S}_a$ denote respectively the sets of all controlled invariant, almost controlled invariant, conditionally invariant, and almost conditionally invariant subspaces then it is possible to show the following closure relations (w.r.t. the Grassmann topology):

$$\underline{V}_a = (\underline{V})^{\text{closure}} \quad \text{and} \quad \underline{S}_a = (\underline{S})^{\text{closure}}$$

These relations show the connection with high gain feedback since they imply that if $V_a \in \underline{V}_a$ but $V_a \notin \underline{V}$ then there will exist $V_\varepsilon \in \underline{V}$ and $F_\varepsilon$ such that $V_\varepsilon \underset{\varepsilon \to 0}{\to} V_a$, $(A + BF_\varepsilon) V_\varepsilon \subset V_\varepsilon$ , and we will necessarily have $F_\varepsilon \xrightarrow[\varepsilon \to 0]{} \infty$. Similarly if $S_a \in \underline{S}_a$ but $S_a \notin \underline{S}$ then the gains K,L of the observer in the definition of $S_a$ will approach infinity as $\varepsilon \to 0$.

The above facts are proven in detail in [1,2], where we give also feedback and output injection characterizations of these subspaces, and relations with controlled invariant subspaces using distributional controls $\underline{u}$ or with conditional invariance with observers which employ differentiations of the observations $\underline{y}$.

The most immediate application of these ideas is to the disturbance decoupling problem. Crucial in this and other applications in this area is the fact that for any subspaces $K$ and $L$ of $X$

$\sup\{V_a \in \underline{V}_a \mid V_a \subset K\} =: V^*_{a,K}$ belongs to $\underline{V}_a$

and $\inf\{S_a \in \underline{S}_a \mid S_a \supset L\} =: S^*_{a,L}$ belongs to $\underline{S}_a$

Let $V^*_{b,K} := V^*_K + AV^*_{a,K} + \mathrm{im}\,B$ and $S^*_{b,L} := S^*_L \cap (A^{-1}S^*_{a,L}) \cap \ker C$ . These subspaces play a crucial role in applications. It can be shown that they are the $L_p$-versions $(1 \leq p < \infty)$ of $V^*_{a,K}$ and $S^*_{a,L}$, which were defined in terms of $L_\infty$-norms. In [1,2] finite recursive algorithms are given for computing all these spaces.

Consider now the disturbance decoupling problem for the plant

$$\Sigma:\ \dot{x} = Ax + Bu + Gd\ ;\ y = Cx\ ;\ z = Hx\ ,$$

with u the control, d the disturbance, y the measurement, and z the to-be-controlled output. The feedback processor is assumed to be given by the finite dimensional linear time-invariant system

$$\Sigma_{fb}:\ \dot{w} = Kw + Ly\ ;\ u = Mw + Fy\ .$$

The closed loop system is now of the form

$$\Sigma_{cl}:\ \dot{x}_e = A_e x_e + G_e d\ ;\ z = H_e x_e\ ,$$

with $A_e, G_e$, and $H_e$ matrices which are easily obtained in terms of (A,B,G,C,H) and (K,L,M,F). In the *approximate disturbance decoupling problem* (ADDPM) we look for conditions on $\Sigma$ such that $\forall \varepsilon > 0\ \exists\ \Sigma_{fb}$ such that the closed loop impulse response of $\Sigma_{cl}$ satisfies

$$\int_\infty^0 \| H_e e^{A_e t} G_e \|\ dt \leq \varepsilon\ .$$

It turns out that this question can be answered very nicely in terms of almost invariant subspaces. In fact:

<u>Theorem:</u> $\{$(ADDPM) *is solvable*$\} \Leftrightarrow \{\mathrm{im}\,G \subset V^*_{b,\ker H}$ *and* $S^*_{b,\mathrm{im}\,G} \subset \ker H\}$

It is interesting to compare this to the conditions for exact disturbance decoupling [4,5], which are $\mathcal{S}^*_{\mathrm{im}\,G} \subset \mathcal{V}^*_{\ker H}$. This condition is quite a bit more restrictive than the one obtained in the above theorem. This has important consequences and yields a number of explicit conditions for solvability of (ADDPM) some of them simply in terms of the numbers of controls, disturbances, measurements, and to-be-controlled outputs.

An immediate consequence of this theorem is the following *separation principle*: approximate disturbance decoupling by measurement feedback is possible if and only if (i) approximate disturbance decoupled control of the to-be-controlled output is possible by means of state feedback and (ii) approximate disturbance decoupled estimation of the to-be-controlled output is possible by means of the measurement.

One of the by-products of the theory beyond the proof of the above theorem is a rather complete geometric theory for the solvability of the following equations

$$G_1(s)X_1(s) = G_2(s) \;,\; X_2(s)G_1(s) = G_2(s) \;,\; \text{and } G_1(s)X_3(s)G_2(s) = G_3(s)$$

with $G_1, G_2, G_3$ (strictly) proper rational matrices and $X_1, X_2$ and $X_3$ unknown matrices which may be rational, (strictly) proper rational, or polynomial matrices depending on the case.

This theory is furthermore immediately related to feedforward control [6] and also allows interpretations for discrete time systems. In [7] it has been applied to non-interacting control synthesis. A number of open problems in this area have been described in [8]. One such open question is to obtain these results from a singular perturbation point of view and to analyse the time-scale behavior of high gain feedback systems in terms of almost invariant subspaces.

Références:

1 J.C. Willems, "Almost invariant subspaces: An approach to high gain feedback design - Part I: Almost controlled invariant subspaces", *IEEE Transactions Automatic Control*, vol. AC-26, pp. 235-252, Feb. 1981.

2 J.C. Willems, "Almost invariant subspaces: An approach to high gain feedback design - Part II: Almost conditionally invariant subspaces", *IEEE Transactions Automatic Control*, vol. AC-27, Oct. 1982

3 W.M. Wonham, *Linear Multivariable Control: A Geometric Approach*, 2nd ed., New York: Springer-Verlag, 1979.

4 J.M. Schumacher, "Compensator design using (C,A,B)-pairs", *IEEE Transactions Automatic Control*, vol. AC-25, pp. 1133-1138, 1980.

5 H. Ahashi and H. Imai, "Disturbance localization and output deadbeat control through an observer in discrete-time linear multivariable systems", *IEEE Transactions Automatic Control*, vol. AC-24, pp. 621-627, 1979.

6 J.C. Willems, "Feedforward control, PID control laws, and almost invariant subspaces", *Systems & Control Letters*, vol. 1, no. 4, pp. 277-282, 1982.

7 J.C. Willems, "Almost noninteracting control design using dynamic state feedback", in *Analysis and Optimization of Systems*, A. Bensoussan and J.L. Lions, Eds., Lecture Notes in Control and Information Sciences, vol. 28, Springer Verlag, pp. 555-561, 1980.

8 J.C. Willems, "Les espaces presque-invariants" in *Méthodes de Perturbation en Automatique*, Editions CNRS, Paris, France, to appear.

# LINEAR QUADRATIC GAUSSIAN ESTIMATION AND CONTROL OF SINGULARLY PERTURBED SYSTEMS

Hassan K. Khalil
Department of Electrical Engineering
& Systems Science
Michigan State University
East Lansing, MI 48824-1226
U.S.A.

Abstract

Techniques are presented for approximating optimal steady-state solutions of time-invariant-linear-quadratic-Gaussian estimation and control problems for linear singularly perturbed systems.

## 1. Introduction

Optimal estimation and control problems for linear singularly perturbed systems have been studied in [1-8]. Haddad gave a near-optimal estimation scheme [1] in which the Kalman filter is decomposed into two lower order Kalman filters for the slow and fast dynamics. Haddad and Kokotovic [2] extended the composite control idea from the deterministic linear quadratic problem [9] to the stochastic linear quadratic Gaussian problem. Teneketzis and Sandell [3] adopted a different approach for approximating optimal controllers by manipulating the equations describing the closed-loop systems. Recently, Khalil and Gajic [8] developed a new approach for approximating singularly perturbed estimation and control problems. The new approach has led to results that extend over the results of Haddad and Kokotovic [1,2]. The other papers [4-7] dealt either with special cases or with cases when the white noise input is appropriately scaled such that the fast variables behave smoothly as the perturbation parameter tends to zero. This lecture will deal with the standard problem formulation as in [1,2,3,8]. The presentation will follow [1], [2] and [8] with emphasis on [8] since the results of [8] incorporate those of [1] and [2] for the case of time invariant problems which is the subject of this lecture. The lecture is organized as follows. Section 2 presents a general approximation result for singularly perturbed systems driven by white noise. This result is used to prove the estimation and control results of the following sections. The estimation problem is considered in section 3 and the control problem is in section 4. Section 5 presents a composite control scheme which is derived from slow and fast decompositions of the problem.

## 2. Approximations of singularly perturbed systems driven by white Noise

Consider the linear time-invariant singularly perturbed system

$$\dot{x}(t) = A(\mu)x(t) + B(\mu)y(t) + E(\mu)w(t), \quad x(0) = x^0(\mu), \tag{1}$$

$$\mu\dot{y}(t) = C(\mu)x(t) + D(\mu)y(t) + F(\mu)w(t), \quad y(0) = y^0(\mu), \tag{2}$$

where $x \in R^n$, $y \in R^m$, $w \in R^r$ and $\mu$ is a small positive scalar parameter. The system matrices are analytic functions in $\mu$ at $\mu = 0$. The input w(t) is zero-mean, stationary, white Gaussian noise with intensity matrix $V > 0$, i.e.,

$$E\{w(t)w^T(s)\} = V\,(t - s). \tag{3}$$

The initial conditons $x^0(\mu)$ and $y^0(\mu)$ are Gaussian random vectors with means $\overline{x}^0(\mu)$ and $\overline{y}^0(\mu)$, and joint variance matrix $\Gamma^0(\mu)$, where $\overline{x}^0(\mu)$, $\overline{y}^0(\mu)$ and $\Gamma^0(\mu)$ are analytic in $\mu$ at $\mu = 0$. It is assumed that the matrices $(A_0 - B_0D_0^{-1}C_0)$ and $D_0$ are Hurwitz, i.e.,

$$\mathrm{Re}\ \lambda(D_0) < 0 \tag{4}$$

and

$$\mathrm{Re}\ \lambda(A_0 - B_0D_0^{-1}C_0) < 0. \tag{5}$$

Conditions (4) and (5) guarantee that for sufficiently small $\mu$ the singularly perturbed system is asymptotically stable. Suppose now that the stochastic processes x(t) and y(t) are approximated by $x_N(t)$ $y_N(t)$ which satisfy a perturbed version of (1) and (2) obtained by making $O(\mu^N)$ perturbations in the matrix coefficients and inital conditions, i.e., $x_N(t)$ and $y_N(t)$ satisfy

$$\dot{x}_N(t) = A^N(\mu)x_N(t) + B^N(\mu)y_N(t) + E^N(\mu)w(t), \quad x_N(0) = x_N^0(\mu), \tag{6}$$

$$\mu\dot{y}_N(t) = C^N(\mu)x_N(t) + D^N(\mu)y_N(t) + F^N(\mu)w(t), \quad y_N(0) = y_N^0(\mu). \tag{7}$$

where the matrices $A^N(\mu)$ through $F^N(\mu)$ are analytic functions in $\mu$ which are $O(\mu^N)$ close to the corresponding matrices $A(\mu)$ through $F(\mu)$, respectively, e.g.,

$$A^N(\mu) - A(\mu) = O(\mu^N), \tag{8}$$

and the initial conditions $x_N^0(\mu)$ and $y_N^0(\mu)$ are Gaussian random vectors which are $O(\mu^N)$ close to $x^0(\mu)$ and $y^0(\mu)$ in the mean square sense, i.e.,

$$E\left\{\begin{pmatrix} x_N^0(\mu) - x^0(\mu) \\ y_N^0(\mu) - y^0(\mu)\end{pmatrix}\begin{pmatrix} x_N^0(\mu) - x^0(\mu) \\ y_N^0(\mu) - y^0(\mu)\end{pmatrix}^T\right\} = O(\mu^{2N}). \tag{9}$$

The validity of such approximations follows from the following theorem.

Theorem 1: For all t $(0 \leq t < \infty)$

$$\mathrm{Var}(x(t) - x_N(t)) = O(\mu^{2N}), \tag{10}$$

$$\mathrm{Var}(y(t) - y_N(t)) = O(\mu^{2N-1}) \tag{11}$$

and

$$E\{(x(t) - x_N(t))(y(t) - y_N(t))^T\} = O(\mu^{2N}). \tag{12}$$

Furthermore, if the initial condition closeness assumption (9) is not satisfied, (10)-(12) hold at steady-state, i.e., as $t \to \infty$. Theorem 1 is proved in [8]. The idea of the proof is to show that the variance of the error satisfies a deterministic singularly perturbed Lyapunov equation which can be easily handled using deterministic singular perturbation techniques. It is emphasized that Theorem 1 holds even though the variances of $y(t)$ and $y_N(t)$ could be $O(\frac{1}{\mu})$ because of the presence of white noise input multiplied by $1/\mu$. The significance of Theorem 1 is that it sets a guideline for approximating solutions of estimation and control problems for linear singularly perturbed systems. According to that guideline one should approximate an exact solution in such a way that when both the exact and approximate systems are represented as systems driven by white noise, their coefficients and initial conditions are $O(\mu^N)$ apart, where the choise of N depends on the desired accuracy. This idea is behind the approximate schemes of the following sections.

## 3. State estimation

Consider the linear time-invariant singularly perturbed state equation

$$\dot{x}_1(t) = A_{11}x_1(t) + A_{12}x_2(t) + G_1w_1(t) \qquad x_1 \varepsilon R^{n_1},\ w_1 \varepsilon R^r, \tag{13}$$

$$\mu\dot{x}_2(t) = A_{21}x_1(t) + A_{22}x_2(t) + G_2w_1(t), \qquad x_2 \varepsilon R^{n_2} \tag{14}$$

together with the observed output

$$y(t) = C_1x_1(t) + C_2x_2(t) + w_2(t), \qquad y \varepsilon R^{\ell}, \tag{15}$$

where $w_1$ and $w_2$ are independent zero-mean stationary white Gaussian noise processes with intensities $V_1 > 0$ and $V_2 > 0$, respectively. A steady-state optimal observer or Kalman filter for (13), (14) is given by [10]

$$\dot{\hat{x}}(t) = A\,\hat{x}(t) + K[y(t) - C\,\hat{x}(t)]; \tag{16}$$

$$K = QC^T V_2^{-1}; \tag{17}$$

Q is the stablizing solution of the algraic Riccati equation

$$0 = AQ + QA^T + GV_1G^T - QC^TV_2^{-1}CQ, \tag{18}$$

where

$$A = \begin{bmatrix} A_{11} & A_{12} \\ \frac{1}{\mu}A_{21} & \frac{1}{\mu}A_{22} \end{bmatrix}, \quad G = \begin{bmatrix} G_1 \\ \frac{1}{\mu}G_2 \end{bmatrix}, \quad C = (C_1, C_2). \tag{19}$$

The properties of the solution of the filter Riccati equation (18) for small $\mu$ can be obtained by dualizing those of the regulator Riccati equation [9,11]. The following lemma summarizes those properties. <u>Lemma 1</u>: Assume that

(A1) $A_{22}$ is nonsingular

(A2) The triple $(A_o, G_o, C_o)$ is stabilizable-detectable, where

$$A_o = A_{11} - A_{12}A_{22}^{-1}A_{21}, \quad G_o = G_1 - A_{12}A_{22}^{-1}G_2,$$

$$C_o = G_1 - C_2A_{22}^{-1}A_{21}.$$

(A3) The triple $(A_{22}, G_2, C_2)$ is stabilizable-detectable. Then the stablizing solution of (18) possesses a power series expansion at $\mu = 0$, i.e.,

$$Q = \sum_{i=0}^{\infty} \frac{\mu^i}{i!} \begin{bmatrix} Q_1^{(i)} & Q_{12}^{(i)} \\ Q_{12}^T(i) & \frac{1}{\mu}Q_2^{(i)} \end{bmatrix}. \tag{20}$$

Furthermore, the matrices $Q_1^{(0)}$, $Q_{12}^{(0)}$, and $Q_2^{(0)}$ are given by

$$Q_1^{(0)} = Q_0,\ Q_{12}^{(0)} = Q_m,\ Q_2^{(0)} = Q_{22}, \tag{21}$$

where

$$Q_m = [Q_0(C_1^T V_2^{-1} C_2 Q_{22} - A_{21}^T) - (A_{12}Q_{22} + G_1 V_1 G_2^T)](A_{22}^T - C_2^T V_2^{-1} C_2 Q_{22})^{-1} \tag{22}$$

and $Q_0$ and $Q_{22}$ are, respectively, the stablizing solutions of the reduced order Riccati equations:

$$0 = (A_0 - G_0 V_1 H_0^T V_0^{-1} C_0) Q_0 + Q_0 (A_0 \;\; G_0 V_1 H_0^T V_0^{1} C_0)^T$$
$$+ G_0 (V_1 - V_1 H_0^T V_0^{-1} H_0 V_1) G_0^T - Q_0 C_0^T V_0^{-1} C_0 Q_0, \tag{23}$$

where

$$H_0 = -C_2 A_{22}^{-1} G_2,\ V_0 = V_2 + D_0 V_1 D_0^T,$$

and

$$0 = A_{22}Q_{22} + Q_{22}A_{22}^T + G_2 V_1 G_2^T - Q_{22} C_2^T V_2^{-1} C_2 Q_{22}. \tag{24}$$

Using Lemma 1, one can approximate the optimal filter gain K by an approximate gain $\underline{K}$ and then implement the filter equation (16). This would lead to reduction in the off-line computations. However, because of the slow-fast nature of the state variables we seek an approximation that reduces the on-line as well as the off-line computations. The reduction in the on-line computations would result from replacing the full-order Kalman filter equation (16) by two lower-order filters that are implemented in different time scales. We are going to propose an approximation which is based on a linear transformation that was introduced in [12] to block diagonalize singularly perturbed systems. The Kalman filter (16) can be rewritten as

$$\dot{\hat{x}}_1(t) = (A_{11} - K_1C_1)\hat{x}_1(t) + (A_{12} - K_1C_2)\hat{x}_2(t) + K_1y(t) \tag{25}$$

$$\mu\dot{\hat{x}}_2(t) = (A_{21} - K_2C_1)\hat{x}_1(t) + (A_{22} - K_2C_2)\hat{x}_2(t) + K_2y(t). \tag{26}$$

where $K_1$ and $K_2$ are defined as

$$K = \begin{bmatrix} K_1 \\ \frac{K_2}{\mu} \end{bmatrix} = \begin{bmatrix} Q_1 & Q_{12} \\ Q_{12}^T & \frac{1}{\mu}Q_2 \end{bmatrix} \begin{bmatrix} C_1^T \\ C_2^T \end{bmatrix} V_2^{-1} \tag{27}$$

We use a transformation to block diagonalize the homogeneous part of (25) and (26). The transformation is

$$\begin{bmatrix} \hat{\eta}_1(t) \\ \hat{\eta}_2(t) \end{bmatrix} = \begin{bmatrix} I_n - \mu M(\mu)L(\mu) & -\mu M(\mu) \\ L(\mu) & I_m \end{bmatrix} \begin{bmatrix} \hat{x}_1 \\ \hat{x}_2 \end{bmatrix} \tag{28}$$

where the matrices L and M are chosen to satisfy the equations

$$0 = (A_{22} - K_2C_2)L - (A_{21} - K_2C_1) - \mu L[(A_{11} - K_1C_1)$$

$$- (A_{12} - K_1C_2)L], \tag{29}$$

$$0 = -M(A_{22} - K_2C_2) + (A_{12} - K_1C_2) - \mu ML(A_{12} - K_1C_2)$$

$$+ \mu[(A_{11} - K_1C_1) - (A_{12} - K_1C_2)L]M. \tag{30}$$

We notice that by Lemma 1 the matrix $(A_{22} - K_2C_2)$ is nonsingular at $\mu = 0$, therefore it follows from [13] that there is $\mu^* > 0$ such that $\forall\mu\epsilon(0,\mu^*)$ there exist L and M matrices satisfying (29) and (30). In fact L and M are analytic in $\mu$ at $\mu = 0$ with

$$L(o) = (A_{22} - K_2(o)C_2)^{-1} (A_{21} - K_2(o)C_1), \tag{31}$$

and

$$M(o) = (A_{12} - K_1(o)C_2) (A_{22} - K_2(o)C_2)^{-1}. \tag{32}$$

The transformed optimal filter is given by

$$\dot{\hat{\eta}}_1 = [(A_{11} - K_1C_1) - (A_{12} - K_1C_2)L]\hat{\eta}_1 + [K_1 - MK_2 - \mu MLK_i]y, \tag{33}$$

$$\mu\dot{\hat{\eta}}_2 = [(A_{22} - K_2C_2) + \mu L(A_{12} - K_1C_2)]\hat{\eta}_2 + [K_2 + \mu LK_1]y, \tag{34}$$

The optimal filter (33), (34) has separate slow and fast parts which can be implemented in separate time scales leading to a reduction in the on-line processing time. To achieve this reduction we did not need to employ any approximations. We have just employed the singularly perturbed nature of the Kalman filter to transform it into new coordinates where it is decomposed into slow and fast parts. The price of this reduction in the on-line processing time is an increase in the off-line computations since two additional matrix equations for L and M have to be solved and the coefficients of (33) and (34) have to be computed. The off-line computations, however, can be reduced by using approximations of $K_1$, $K_2$, L and M. We know that these four matrices are analytic in $\mu$ at $\mu = 0$. Let $K_1^N$, $K_2^N$, $L^N$ and $M^N$ be Nth-order approximations of $K_1$, $K_2$, L and M, respectively, e.g.,

$$K_1(\mu) - K_1^N(\mu) = O(\mu^N). \tag{35}$$

Such approximations can be obtained using one of two methods. The first method is based on generating power expansions of these matrices. The second one is based on iterative solutions of the Riccati equation and the equations (29), (30) for L and M. The details of the first method are given in [8]. The basic idea is to write a power series expansion in $\mu$ for each one of the matrices under consideration. Those expansions are substituted in the filter Riccati equation and equations (29), (30). By matching coefficients of equal powers of $\mu$, one can obtain equations which uniquely define the coefficients of the power series expansions. The Nth-order approximation of a matrix is then taken as the N leading terms of the respective expansion. Although this method is theoretically sound its numerical implementation is inferior to the second method. So the second method will be given in more detail.

The second method is based on solving the filter Riccati equation (18) and equations (29) and (30) iteratively using successive approximations. One starts with the filter Riccati equation (18). We know that

$$Q_1(0) = Q_0, \quad Q_{12}(0) = Q_m, \quad Q_2(0) = Q_{22}$$

where $Q_0$, $Q_m$ and $Q_{22}$ are given by (22)-(24). Let

$$Q_i(\mu) = Q_i(0) + \mu E_i(\mu), \quad i = 1,2$$

$$Q_{12}(\mu) = Q_{12}(0) + \mu E_{12}(\mu).$$

The matrices $E_1(\mu)$, $E_{12}(\mu)$ and $E_2(\mu)$ satisfy the following equations

$$F_1 E_1(\mu) + E_1(\mu) F_1^T + F_2 E_{12}(\mu) + E_{12}^T(\mu) F_2^T = \mu f_1(E_1(\mu), E_{12}(\mu),\mu) \tag{36}$$

$$E_1(\mu) F_3^T + E_{12}^T(\mu) F_4^T + F_2 E_2(\mu) = R_2 + \mu f_2(E_1(\mu), E_{12}(\mu), E_2(\mu),\mu) \tag{37}$$

$$F_4 E_2(\mu) + E_2(\mu) F_4^T = R_3 + \mu f_3(E_{12}(\mu), E_2(\mu),\mu) \tag{38}$$

The matrices $F_1$, $F_2$, $F_3$, $F_4$, $R_2$ and $R_3$ are independent of $\mu$ and the nonlinear functions $f_1$, $f_2$ and $f_3$ are analytic in $\mu$, $E_1$, $E_{12}$ and $E_2$. The expressions for these quantities, together with the details of this iterative method, will be given elsewhere. For our purpose here it is enough to know that for any given righthand side, equations (36)-(38) have a unique solution for $E_1$, $E_{12}$ and $E_2$. Based on this observation the solution of (36)-(38) can be sought iteratively using successive approximations. The iterations are initiated at $E_1 = 0$, $E_{12} = 0$, and $E_2 = 0$. It can be shown that for sufficiently small $\mu$ these iterations converge to the exact solution. Furthermore, if $E_1^N$, $E_{12}^N$ and $E_2^N$ are the solutions obtained in the Nth-iteration and if $Q_1^N$, $Q_{12}^N$ and $Q_2^N$ are taken as

$$Q_i^N = Q_i(0) + \mu E_i^{N-1}, \qquad i = 1,2$$

$$Q_{12}^N = Q_{12}(0) + \mu E_{12}^{N-1},$$

and $K_1^N$, $K_2^N$ are taken as

$$\begin{bmatrix} K_1^N \\ \frac{1}{\mu}K_2^N \end{bmatrix} = \begin{bmatrix} Q_1^N & Q_{12}^N \\ Q_{12}^{NT} & \frac{1}{\mu}Q_2^N \end{bmatrix} \begin{bmatrix} C_1^T \\ C_2^T \end{bmatrix} V_2^{-1}$$

then

$$\| K_i(\mu) - K_i^N(\mu) \| = O(\mu^N), \qquad i = 1,2.$$

After computing $K_1^N$ and $K_2^N$ one can seek the solution of (29) and (30) iteratively using the method of [13]. In performing those iterations $K_1$ and $K_2$ in (29), (30) are fixed at $K_1^N$ and $K_2^N$. If $L^N$ and $M^N$ are the solutions obtained in the Nth-iteration, then

$$\| L(\mu) - L^N(\mu) \| = O(\mu^N) \text{ and } \| M(\mu) - M^N(\mu) \| = O(\mu^N).$$

Now, irrespective of the method used to compute the Nth-order approximations $K_1^N$, $K_2^N$, $L^N$ and $M^N$, once these matrices are available, approximate estimates of $x_1$ and $x_2$ are obtained by implementing the approximate filter.

$$\dot{\hat{\eta}}_1^N(t) = [(A_{11} - K_1^N C_1) - (A_{12} - K_1^N C_2)L^N]\hat{\eta}_1^N(t)$$

$$+ [K_1^N - M^N K_2^N - \mu M^N L^N K_1^N]y(t), \quad (39)$$

$$\mu\dot{\hat{\eta}}_2^N(t) = [(A_{22} - K_2^N C_2) + \mu L^N(A_{12} - K_1^N C_2)]\hat{\eta}_2^N(t)$$

$$+ [K_2^N + \mu L^N K_1^N]y(t), \quad (40)$$

$$\hat{x}_1^N(t) = \hat{\eta}_1^N(t) + \mu M^N \hat{\eta}_2^N(t), \quad (41)$$

$$\hat{x}_2^N(t) = -L^N \hat{\eta}_1^N(t) + (I_m - \mu L^N M^N)\hat{\eta}_2^N(t). \quad (42)$$

These approximate estimates are justified by the following theorem

Theorem 2 : If $R_e\ \lambda(A_0) < 0$ and $Re\ \lambda(A_{22}) < 0$, then the following relations hold at steady-state (as $t \to \infty$):

$$Var(\hat{x}_1(t) - \hat{x}_1^N(t)) = 0(\mu^{2N}) \quad (43)$$

and

$$Var(\hat{x}_2(t) - \hat{x}_2^N(t)) = 0(\mu^{2N-1}). \quad (44)$$

Theorem 2 is proved in [8]. The proof applies Theorem 1.

## 4. Linear-Quadratic-Gaussian control

Consider the singularly perturbed system

$$\dot{x}_1(t) = A_{11}x_1(t) + A_{12}x_2(t) + B_1u(t) + G_1w_1(t), \quad (45)$$

$$\mu\dot{x}_2(t) = A_{21}x_1(t) + A_{22}x_2(t) + B_2u(t) + G_2w_1(t), \quad (46)$$

$$y(t) = C_1x_1(t) + C_2x_2(t) + w_2(t) \quad (47)$$

with the performance criterion

$$J = \lim_{\substack{t_0 \to -\infty \\ t_1 \to \infty}} \frac{1}{t_1 - t_0} E\left\{\int_{t_0}^{t_1} [z^T(t)z(t) + u^T(t)R\,u(t)]dt\right\} \tag{48}$$

where $x_1$, $x_2$, $w_1$, $w_2$ and $y$ are as defined in section 3, $u \epsilon R^m$ is the control variable and $z \epsilon R^s$ is the controlled output, which is given by

$$z(t) = D_1 x_1(t) + D_2 x_2(t).$$

The optimal control law is given by [2,10]

$$\dot{\hat{x}}_1(t) = A_{11}\hat{x}_1(t) + A_{12}\hat{x}_2(t) + B_1 u(t) + K_1(\mu)[y(t) - C_1\hat{x}_1(t) - C_2\hat{x}_2(t)], \tag{49}$$

$$\mu\dot{\hat{x}}_2(t) = A_{21}\hat{x}_1(t) + A_{22}\hat{x}_2(t) + B_2 u(t) + K_2(\mu)[y(t) - C_1\hat{x}_1(t) - C_2\hat{x}_2(t)], \tag{50}$$

$$u(t) = -[F_1(\mu)\hat{x}_1(t) + F_2(\mu)\hat{x}_2(t)]. \tag{51}$$

The filter gains $K_1$ and $K_2$ have been already given in section 3. The regulator gains $F_1$ and $F_2$ are given by

$$(F_1, F_2) = R^{-1}(B_1^T \;\; \tfrac{1}{\mu}B_2^T) \begin{bmatrix} P_1 & \mu P_{12} \\ \mu P_{12}^T & \mu P_2 \end{bmatrix} \triangleq R^{-1}B^T P \tag{52}$$

where P is the stablizing solution of the algebraic Riccati equation

$$0 = PA + A^T P + D^T D - PBR^{-1}B^T P \tag{53}$$

with $D = (D_1, D_2)$

The following lemma, which is recalled from [9], is the dual of Lemma 1.

Lemma 2: Assume that

(A1) $A_{22}$ is nonsingular

(A2) The triple $(A_0, B_0, D_0)$ is stabilizable-detectable, where

$$B_0 = B_1 - A_{12}A_{22}^{-1}B_2, \quad D_0 = D_1 - D_2A_{22}^{-1}A_{21}.$$

(A3) The triple $(A_{22}, B_2, D_2)$ is stabilizable-detectable.

Then the stablizing solution of (53) possesses a power series expansion at $\mu = 0$, i.e.,

$$P = \sum_{i=0}^{\infty} \frac{\mu^i}{i!} \begin{bmatrix} P_1^{(i)} & \mu P_{12}^{(i)} \\ \mu P_{12}^{T(i)} & \mu P_2^{(i)} \end{bmatrix} \tag{54}$$

Furthermore

$$P_1^{(0)} = P_0, \quad P_{12}^{(0)} = P_m, \quad P_2^{(0)} = P_{22}, \tag{55}$$

where

$$P_m = [P_0(B_1R^{-1}B_2^TP_{22} - A_{12}) - (A_{21}^TP_{22} + D_1^TD_2)](A_{22} - B_2R^{-1}B_2^TP_{22})^{-1} \tag{56}$$

and $P_0$ and $P_{22}$ are, respectively, the stablizing solutions of the reduced order Riccati equations

$$\begin{aligned} 0 = {} & P_0(A_0 - B_0R_0^{-1}E_0^TD_0) + (A_0 - B_0R_0^{-1}E_o^TD_o)^TP_o \\ & + D_o^T(I - E_0R_0^{-1}E_0^T)D_0 - P_0B_0R_0^{-1}B_o^TP_o \end{aligned} \tag{57}$$

where

$$E_0 = -D_2A_{22}^{-1}B_2, \quad R_0 = R + E_0^TE_0$$

and

$$0 = P_{22}A_{22} + A_{22}^T P_{22} + D_2^T D_2 - P_{22}B_2R^{-1}B^T P_{22}. \tag{58}$$

A near-optimal control is derived in the spirit of the separation principle. For any input u, the near-optimal filter of section 3 is used to approximate $\hat{x}_1$ and $\hat{x}_2$. With the estimates replacing the actual states, the near-optimal deterministic state feedback control law of [9,11] is used. The resulting approximate control law is given by

$$\dot{\hat{\eta}}_1^N(t) = [(A_{11} - K_1^N C_1) - (A_{12} - K_1^N C_2)L^N]\hat{\eta}_1^N(t)$$
$$+[K_1^N - M^N K_2^N - \mu M^N L^N K_1^N]y(t) + [B_1 - M^N B_2 - \mu M^N L^N B_1]u(t) \tag{59}$$

$$\mu\dot{\hat{\eta}}_2^N(t) = [(A_{22} - K_2^N C_2) + \mu L^N(A_{12} - K_1^N C_2)]\hat{\eta}_2^N + [K_2^N + \mu L^N K_1^N]y(t)$$
$$+[B_2 + \mu L^N B_1]u(t), \tag{60}$$

$$\hat{x}_1^N(t) = \hat{\eta}_1^N(t) + \mu M^N\hat{\eta}_2^N(t), \tag{61}$$

$$\hat{x}_2^N(t) = -L^N\hat{\eta}_1^N(t) + (I_m - \mu L^N M^N)\hat{\eta}_2^N(t), \tag{62}$$

$$u(t) = -F_1^N\hat{x}_1^N(t) - F_2^N\hat{x}_2^N(t) \tag{63}$$

Equations (59)-(63) are obtained by using the state transformation (28) to block diagonalize the homogeneous part of the Kalman filter (49), (50) with u and y being treated as deriving inputs; then, $K_1$, $K_2$, $F_1$, $F_2$, L and M are approximated by their Nth-order approximations $K_1^N$, $K_2^N$, $F_1^N$, $F_2^N$, $L^N$ and $M^N$. The matrices $K_1^N$, $K_2^N$, $L^N$ and $M^N$ are obtained as in section 3 while $F_1^N$ and $F_2^N$ are obtained similarly by using the power series expansion method or the iterative method when applied to the regulator Riccati equation (53). Theorem 3 below, which is proved in [8], shows the near-optimality of the control law (59)-(63).

Theorem 3: Suppose that the conditions of Lemma 1 and Lemma 2 hold. Let $x_1^*(t)$ and $x_2^*(t)$ be the optimal trajectories and $J^*$ be the optimal value of the performance criterion. Let $x_1(t)$, $x_2(t)$ and J be the corresponding quantities under the control law (59)-(63); and let $\Delta J = J - J^*$.

Then

$$\frac{\Delta J}{J^*} = O(\mu^N), \tag{64}$$

$$Var(x_1(t) - x_1^*(t)) = O(\mu^{2N}), \qquad (\text{as } t \to \infty), \tag{65}$$

and

$$Var(x_2(t) - x_2^*(t)) = O(\mu^{2N-1}), \qquad (\text{as } t \to \infty). \tag{66}$$

5. Two stage design

A near-optimal control law will be derived based on intuitive slow-fast decompositions of the system. We start by employing the separation principle to replace the optimal control problem (45)-(48) by separate estimation and regulation problems. The regulation problem is defined by

$$\dot{x}_1(t) = A_{11}\, x_1(t) + A_{12}\, x_2(t) + B_1\, u(t) \tag{67}$$

$$\mu\dot{x}_2(t) = A_{21}\, x_1(t) + A_{22}\, x_2(t) + B_2\, u(t) \tag{68}$$

$$z(t) = D_1\, x_1(t) + D_2\, x_2(t) \tag{69}$$

$$J = \int_0^\infty [z^T(t)\, z(t) + u^T(t) R\, u(t)] dt \tag{70}$$

where it is assumed that $x_1(t)$ and $x_2(t)$ are perfectly measureable. A two-stage state feedback control law has been derived in [9] by solving separate slow and fast control problems. The slow control problem is defined by neglecting fast transients. This is "formally" equivalent to setting $\mu = 0$ in the L.H.S. of (68) and eliminating $x_2$. The resulting slow control problem is given by

$$\dot{x}_s(t) = A_0\, x_s(t) + B_0\, u_s(t), \tag{71}$$

$$J_s = \int_0^\infty [x_s^T(t)\, D_0^T D_0\, x_s(t) + 2x_s^T(t)\, E_0^T D_0\, x_s(t) + u_s^T(t) R_0\, u_s(t)] dt \tag{72}$$

and its optimal solution, which is guaranteed to exist under the conditions of Lemma 2, is given by

$$u_s(t) = -R_0^{-1}(E_0^T D_0 + B_0^T P_0)\, x_s(t) \triangleq -F_0\, x_s(t) \tag{73}$$

The fast control problem is defined by assuming that the slow variables are constant during fast transients and by introducing $x_f(t)$ and $u_f(t)$ to represent pure fast transients. The resulting fast control problem is given by

$$\mu \dot{x}_f(t) = A_{22}\, x_f(t) + B_2\, u_f(t), \tag{74}$$

$$J_f = \frac{1}{2} \int_0^\infty [x_f^T(t)\, D_2^T D_2\, x_f(t) + u_f^T(t) R\, u_f(t)] dt \tag{75}$$

and its optimal solution, which is guaranteed to exist under the conditions of Lemma 2, is given by

$$u_f(t) = -R^{-1}B_2^T P_{22}\, x_f(t) \triangleq -F_{22}\, x_f(t). \tag{76}$$

A composite control law $u_c = u_s + u_f$ is then expressed in terms of $x_1$ and $x_2$ as

$$u_c(t) = -F_o\, x_1(t) - F_{22}\, [x_2(t) - A_{22}^{-1}B_2F_o\, x_1(t) + A_{22}^{-1}A_{21}\, x_1(t)]. \tag{77}$$

Next, we consider the estimation problem which is defined by

$$\dot{x}_1(t) = A_{11}\, x_1(t) + A_{12}\, x_2(t) + B_1\, u(t) + G_1\, w_1(t) \tag{78}$$

$$\mu\dot{x}_2(t) = A_{21}\, x_1(t) + A_{22}\, x_2(t) + B_2\, u(t) + G_2\, w_1(t) \tag{79}$$

$$y(t) = C_1\, x_1(t) + C_2\, x_2(t) + w_2(t) \tag{80}$$

A two stage estimator can be designed as follows. First, a slow estimation problem is defined by assuming that in estimating the slow variable $x_1$ the fast transients of $x_2$ can be neglected. This is "formally" equivalent to setting $\mu = 0$ in the R.H.S. of (79) and eliminating $x_2$. The resulting slow estimation problem, for any input $u(t)$, is given by

$$\dot{\bar{x}}_1(t) = A_o\, \bar{x}_1(t) + B_o\, u(t) + G_o\, w_1(t), \tag{81}$$

$$y(t) = C_o\, \bar{x}_1(t) + N_o\, u(t) + H_o\, w_1(t) + w_2(t). \tag{82}$$

The steady-state optimal estimate, denoted by $\eta_1(t)$, exists under the conditions of Lemma 1 and is the output of the Kalman filter

$$\dot{\eta}_1(t) = A_o\ \eta_1(t) + B_o\ u(t) + K_o(y(t) - C_o\ \eta_1(t) - N_o\ u(t)),$$
(83)

where

$$K_o = (Q_o C_o^T + G_o V_1 H_o) V_o^{-1}.$$

Second, a fast estimation problem is defined by assuming that in estimating the fast variable $x_2$ the slow variable $x_1$ can be treated as a constant $\bar{x}_1$, that is,

$$\mu \dot{\bar{x}}_2(t) = A_{21}\ \bar{x}_1 + A_{22}\ \bar{x}_2(t) + B_2\ u(t) + G_2\ w_1(t), \tag{84}$$

$$y(t) = C_1\ \bar{x}_1 + C_2\ \bar{x}_2(t) + w_2(t). \tag{85}$$

The effect of the constant terms on the R.H.S. of (84) and (85) can be handled via the shifting transformation

$$\xi(t) = \bar{x}_2(t) + A_{22}^{-1}\ A_{21}\ \bar{x}_1 \tag{86}$$

$$y_f(t) = y(t) - C_o \bar{x}_1. \tag{87}$$

The estimation problem for $\xi(t)$ is given by

$$\mu \dot{\xi}(t) = A_{22}\ \xi(t) + B_2\ u(t) + G_2\ w_1(t), \tag{88}$$

$$y_f(t) = C_2\ \xi(t) + w_2(t). \tag{89}$$

The steady-state optimal estimate $\hat{\xi}(t)$, which exists under the conditions of Lemma 1, is the output of the Kalman filter

$$\mu \dot{\hat{\xi}}(t) = A_{22}\ \hat{\xi}(t) + B_2\ u(t) + K_{22}\ (y_f(t) - C_2\ \hat{\xi}(t)), \tag{90}$$

where

$$K_{22} = Q_{22}C_2^T V_2^{-1}.$$

In implementing the Kalman filter (90), $y_f(t)$ has to be substituted by $y(t)$ using (87) since $y(t)$ is the physically measurable output The effect of the constant term $C_0\,\bar{x}_1$ at steady-state is to cause the constant bias $-(A_{22} - K_{22}C_2)^{-1}\,K_{22}C_0\,\bar{x}_1$. So, if we let $\eta_2(t)$ be the output of

$$\mu\dot{\eta}_2(t) = A_{22}\,\eta_2(t) + B_2\,u(t) + K_{22}\,(y(t) - C_2\,\eta_2(t)), \tag{91}$$

$\hat{\xi}(t)$ should be taken as

$$\hat{\xi}(t) = \eta_2(t) + (A_{22} - K_{22}C_2)^{-1}\,K_{22}C_0\,\bar{x}_1 \tag{92}$$

Finally, the estimates of $x_1(t)$ and $x_2(t)$ will be approximated by

$$\hat{x}_1(t) = \eta_1(t), \tag{93}$$

$$\hat{x}_2(t) = \eta_2(t) - A_{22}^{-1}\,A_{21}\,\eta_1(t) + (A_{22} - K_{22}\,C_2)^{-1}\,K_{22}\,C_o\,\eta_1(t) \tag{94}$$

where in writing (94) the equations (86) and (92) have been employed with $\bar{x}_1$ replaced by the slow variable estimate $\hat{x}_1(t)$. The composite state feedback control law (77) can now be implemented with $x_1$ and $x_2$ replaced by their estimates $\hat{x}_1$ and $\hat{x}_2$ as given by (93) and (94). The near-optimality of this composite control law is established in the following theorem.

<u>Theorem 4</u>: Suppose that the conditions of Lemma 1 and Lemma 2 hold. Let $x_1^*(t)$ and $x_2^*(t)$ be the optimal trajectories and $J^*$ be the optimal

cost. Let $x_1(t)$, $x_2(t)$ and J be the corresponding quantities under the composite control law. Then

$$\frac{J - J^*}{J^*} = O(\mu), \tag{95}$$

$$Var(x_1(t) - x_1^*(t)) = O(\mu^2) \qquad (\text{as } t \to \infty) \tag{96}$$

and

$$Var(x_2(t) - x_2^*(t)) = O(\mu) \qquad (\text{as } t \to \infty). \tag{97}$$

Theorem 4 is proved in [8]. The idea of the proof is to show that the composite control law is nothing more than a special case of the near-optimal control law (59)-(63) when all coefficient of order $O(\mu)$ are neglected.

References

1. Haddad, A., Linear filtering of singularly perturbed systems, IEEE Trans. Aut. Control, 21, 515, 1976.

2. Haddad, A. and Kokotovic, P., Stochastic control of linear singularly perturbed systems, IEEE Trans. Aut. Control, 22, 815, 1977.

3. Teneketzis, D. and Sandell, N., Linear regulator design for stochastic systems by multiple time-scale method, IEEE Trans. Aut. Control, 22, 615, 1977.

4. Rauch, H.E., Application of singular perturbation to optimal estimation, Proc. 11th Annu. Allerton Conf. Circuit and System Theory, 718, 1973.

5. Haddad, A.H., On singular perturbations in stochastic dynamic systems, Proc. 10th Asilomar Conf. on Circuits, Systems and Computers, 94, 1976.

6. Khalil, H.K., Control of linear singularly perturbed systems with colored noise disturbance, Automatica, 14, 153, 1978.

7. Khalil, H.K., Haddad, A. and Blankenship, G., Parameter scaling and well-posedness of stochastic singularly perturbed control systems, Proc. 12th Asilomar Conf. on Circuits, Systems and Computers, 407, 1978.

8. Khalil, H.K. and Gajic, Z., Near-optimum regulators for stochastic linear singularly perturbed systems, research report, College of Engineering, Michigan State University, 1982.

9. Chow, J.H. and Kokotovic, P.V., A decomposition of near-optimum regulators for systems with slow and fast modes, IEEE Trans. Aut. Control, 21, 701, 1976.

10. Kwakernaak, H. and Sivan, R., Linear Optimal Control Systems, Wiley, Interscience, 1972.

11. Yackel, R.A. and Kokotovic, P.V., A boundary layer method for the matrix Riccati equation, IEEE Trans. Aut. Control, 18, 17, 1973.

12. Chang, K., Singular perturbations of a general boundary value problem, SIAM J. Math. Anal., 3, 520, 1972.

References (con't)

13. Kokotovic, P.V., A Riccati equation for block diagonalization of ill conditioned systems, IEEE Trans. Aut. Control, 20, 812, 1975.

# OPTIMAL CONTROL OF PERTURBED MARKOV CHAINS: THE MULTITIME SCALE CASE.

J.P. QUADRAT

INRIA
Domaine de Voluceau - B.P. 105
78153 LE CHESNAY Cédex

ABSTRACT

Given a controled perturbed Markov chain of transition matrix $m^u(\varepsilon)$, where $\varepsilon$ is the perturbation scale and u the control, we study the solution expansion in $\varepsilon$, $w^\varepsilon$, of the dynamic programming equation :

$$\min_u [m^u(\varepsilon)\, w^\varepsilon + c^u(\varepsilon)] = (1+\lambda(\varepsilon))w^\varepsilon .$$

$m^u(\varepsilon)$, $c^u(\varepsilon)$, $\lambda(\varepsilon)$ are polynomials in $\varepsilon$. The case $\lambda(\varepsilon) = \varepsilon^\ell$ leads to study Markov chains on a time scale of order $1/\varepsilon^\ell$. The state space and the control set are finite.

PLAN

1) Introduction
2) Notations and statement of the problem
3) Review of Markov chains
4) Perturbed Markov chains
5) Review of controlled Markov chains.
6) Control of perturbed Markov chains.

## 1. - INTRODUCTION

Stochastic or deterministic control problems can be reduced after discretization to the control of Markov chains. This approach leads to control of Markov chains which have a large number of states. An attempt to solve this difficulty is to see the initial Markov chains as the perturbation of a simpler one when this is possible. "Simple" Markov chains are Markov chains which have several recurrent classes. Then the perturbation can be seen as a small coupling between these recurrent classes. This coupling cannot be neglected on time scale of order $\frac{t}{\varepsilon}$, $\frac{t}{\varepsilon^2}$, where $\varepsilon$ denotes the amplitude of the perturbations. Nevertheless this point of view leads to a hierarchy of more and more aggregated chains, each one being valid for a particular time scale. Their states are the recurrent classes of the faster time scale and their transition matrices can be computed explicitly. Then, in the control context, we can take advantage of this particular structure to design a faster algorithm to solve the dynamic programming equation.

This kind of problem has a long history. Gauss, for example, has studied such problems in celestian mechanics, there the recurrent classes role are played by the planet orbits. In the operations research literature studies of two time scale Markov chains has been done in Simon-Ando [12], Courtois [4], Gaitsgorï-Pervozvanski [8]. The multitime scale situation can be found in Delebecque [5], Coderch-Sastry-Willsky-Castanon [2],[3]. The two time scale control problem (actualization rate of order $\varepsilon$) is solved in Delebecque-Quadrat [6],[7]. The ergodic control problem when the unperturbed chain has no transient classes has been studied in Philips-Kokotovic [19]. In this paper we give the construction of the complete expansion of the optimal cost of the control problem in the general multi-time scale situation. For that we use three kinds of results :

- the Delebecque [5] result describing the reduction process of Kato [9] in the Markov chain situation.
- the realization theory of implicit systems developed by Bernhard [1]. This gives a recursive mean of computing all the cost expansion in the uncontrolled case.
- the Miller-Veinott [10] way of constructing the optimal cost expansion of an unperturbed Markov chain having a small actualization rate.

## 2. - NOTATIONS AND STATEMENT OF THE PROBLEM

We study the evaluation of a cost associated to the trajectory of a discrete Markov chain in four situations (unperturbed-perturbed), (controled-uncontroled); for this let us introduce some n-tuple defining completely the data of each problem, and some related notations.

2.1. - $(T,\mathcal{X},m,c,\lambda)$ is associated to the unperturbed uncontroled case and shall be called the Markov chain n-uple.

- T is the time set isomorphic to $\mathbb{N}$ ;
- $\mathcal{X}$ is the state space of the Markov chain, is a finite discrete space. $|\mathcal{X}|$ denotes card($\mathcal{X}$) that is the number of elements of $\mathcal{X}$. x will be the generic element of $\mathcal{X}$ ;
- m is the transition matrix of the Markov chain, that is a $(|\mathcal{X}|,|\mathcal{X}|)$-matrix with positive entries such that $\sum_{x'\in\mathcal{X}} m_{x\,x'} = 1$ ;
- c is the instantaneous cost that is a $|\mathcal{X}|$-vector with positive entries;
- $\lambda$ is an actualization rate that is, $\lambda \in R$ and $\lambda > 0$.

The set of possible trajectories is denoted by $\Omega = \mathcal{X}^T$, a trajectory by $\omega \in \Omega$, the position of the process at time t if the trajectory is $\omega$ by $X(t,\omega)$. The conditional probability of the cylinder :

$$B = \{\omega : X_t(\omega) = x_t,\ t = 0,1,\dots,n\}$$

knowing $X(0,\omega)$ is :

$$P^{X_0}(B) = \prod_{t=0}^{n-1} m_{x_t x_{t+1}}$$

To the trajectory $\omega$ is associated the cost :

$$j(\omega) = \sum_{t=0}^{+\infty} \frac{1}{(1+\lambda)^{t+1}} c_{X(t,\omega)} \tag{2.1}$$

The conditional expected cost knowing $X(0,\omega)$ is a $|\mathfrak{X}|$-vector denoted w defined by :

$$w_x := E[j(\omega) \mid X(0,\omega) = x], \forall x \in \mathfrak{X} \tag{2.2}$$

The Hamiltonian is the operator :

$$h : \mathbb{R}^{|\mathfrak{X}|} \to \mathbb{R}^{|\mathfrak{X}|} \quad w \mapsto [m-(1+\lambda)i]w+c \tag{2.3}$$

where i denotes the identity of the set of $(|\mathfrak{X}|,|\mathfrak{X}|)$-matrices.

Then w defined by (2.2) is the unique solution of the Kolmogorov equation :

$$h(w) = 0 \tag{2.4}$$

2.2. - In the perturbed situation the n-tuple defining the perturbed Markov chain is :

$$(T, \mathfrak{X}, \mathcal{E}, m(\varepsilon), c(\varepsilon), \lambda(\varepsilon))$$

- $\mathcal{E}$ is now the space of the perturbations ; in all the following i $\mathbb{R}^+$ ;
- $m(\varepsilon)$, $c(\varepsilon)$, $\lambda(\varepsilon)$ have the same definition as previously but depends on the parameter $\varepsilon \in \mathcal{E}$, and we suppose that they are polynomials in this variable.

We denote by $d°$ the degree of a polynomial and by v its valuation (the smallest non zero power of the polynominal). In the following $d^o(m) = 1$, $v(m)=0$, $v(\lambda) = v(c) = \ell$.

The Hamiltonian of the perturbed problem is denoted by :

$$h(w,\varepsilon) = [m(\varepsilon) - (1+\lambda(\varepsilon))i]W + c(\varepsilon) \tag{2.5}$$

The expected conditional cost is denoted $w^\varepsilon$ and is solution of the Kolmogorov equation :

$$h(w^\varepsilon,\varepsilon) = 0 \tag{2.6}$$

We shall prove that $w^\varepsilon$ admits an expansion in $\varepsilon$ that we shall denote by $W(\varepsilon) = \sum_{n=0}^{\infty} \varepsilon^n w_n$ where $W_i$ are $|\mathfrak{X}|$-vectors; then we have :

$$m(\varepsilon)\ W(\varepsilon) = \sum_{n=0}^{\infty} \varepsilon^n (MW)_n \tag{2.7}$$

with :

$$M = \begin{bmatrix} m_0 & & & 0 \\ m_1 & m_0 & & \\ 0 & m_1 & m_0 & \\ & \ddots & \ddots & \ddots \end{bmatrix} \tag{2.8}$$

an infinite block matrix.

For the Hamiltonian we can introduce the same notation :

$$h(W(\varepsilon),\varepsilon) = \sum_{n}^{\infty} \varepsilon^n H_n(W) \tag{2.9}$$

where $H_n(W)$ are the $|\mathfrak{X}|$-vectors defined in (2.9) by identification of the $\varepsilon^i$ terms; that is :

$$\begin{cases} H_0(W) = (m_0-i)w_0 \\ H_1(W) = m_1\, w_0 + (m_0-i)w_1 \\ \vdots \\ H_\ell(W) = -\lambda_\ell\, w_o + m_1\, w_{\ell-1} + (m_0-i)w_\ell + c_\ell \\ \vdots \end{cases} \tag{2.10}$$

(2.10) can written H(W) with :

$$H(W) \equiv [M - (I+\Lambda)]W + C, \tag{2.11}$$

where :

$C = (c_n,\ n \in \mathbb{N},\ c_n$ are $|\mathcal{X}|$-vectors)

I : the identity operator $\begin{bmatrix} i & 0 & 0 & 0 & \cdots \\ 0 & i & 0 & 0 & \cdots \\ 0 & 0 & i & & \\ \vdots & \vdots & & \ddots & \end{bmatrix}$

$\Lambda$ : the operator $\ell^{th}$ $|\mathcal{X}|$-block $\begin{bmatrix} & & 0 \\ i\lambda_\ell & \ddots & \\ 0 & \ddots & \ddots \end{bmatrix}$

An expansion of the cost is obtained by solving :

$$H(W) = 0 \tag{2.12}$$

Moreover the sequence $(W_i,\ i \in \mathbb{N})$ can be computed recursively. These two results will be shown in part 4.

2.3. - For the control problem we need the introduction of the n-tuple:

$$(T, \mathcal{X}, \mathcal{U},\ m^u,\ c^u,\ \lambda)$$

- $\mathcal{U}$ is the set of control which is here a finite set ; $|\mathcal{U}|$ denotes the cardinal of $\mathcal{U}$ ; its generic element is denoted by u ;
- m denotes the $(|\mathcal{U}|, |\mathcal{X}|, |\mathcal{X}|)$ tensor of entries $m^u_{xx'}$, the probability to

go in x', starting from x, the control being u.

- c denotes the $(|\mathcal{U}|,|\mathcal{X}|)$ matrix of entries $c_x^u$, the cost to be in x, the control being u.

A policy is an application :

$$s : \mathcal{X} \to \mathcal{U} .$$

The set of policies is $\mathcal{P} := \mathcal{U}^{\mathcal{X}}$.

For a policy s, mos denotes the $(|\mathcal{X}|,|\mathcal{X}|)$ transition matrix of entries :

$$(mos)_{xx'} = m_{xx'}^{s_x} ; \tag{2.13}$$

cos denotes the $|\mathcal{X}|$-vector :

$$(cos)_x = c_x^{s_x}. \tag{2.14}$$

We associate to a policy $s \in \mathcal{P}$ and a trajectory $\omega$, the cost

$$j^s(\omega) = \sum_{t=0}^{+\infty} \frac{1}{(1+\lambda)^{t+1}} (cos)_{X(t,\omega)} \tag{2.15}$$

and the optimal conditional expected cost knowing the initial condition is :

$$w_x^* = \operatorname*{Min}_{s\in\mathcal{P}} \mathbb{E}(j^s(\omega) \mid X(0,\omega) = x) \tag{2.16}$$

The Hamiltonian is defined as the operator :

$$\begin{aligned} h : \mathcal{U} \times \mathbb{R}^{\mathcal{X}} &\to \mathbb{R}^{\mathcal{X}} \\ (u,w) \quad & h^u(w) = [m^u-(1+\lambda)i]w+c^u. \end{aligned} \tag{2.17}$$

The notation $(hos)_x$ for $h_x^{s_x}$ will be used.

Then the optimal Hamiltonian is the operator :

$$h^* : \mathbb{R}^{\mathcal{X}} \to \mathbb{R}^{\mathcal{X}}, \quad w \quad h^*_x(w) = \min_u h^u_x(w), \forall x \in \mathcal{X} \tag{2.18}$$

The optimal expected cost $w^*$ is the unique solution of the dynamic programming equation :

$$h^*(w^*) = 0 \tag{2.19}$$

An optimal policy is given by :

$$s^* : \mathcal{X} \to \mathcal{U}, \quad x \quad s^*_x \in \operatorname{argmin} h^u_x(w^*), \forall x \in \mathcal{X}.$$

2.4. - The perturbed control problem is defined by the n-tuple:

$$(T, \mathcal{X}, \mathcal{U}, \mathcal{E}, m^u(\varepsilon), c^u(\varepsilon), \lambda(\varepsilon)).$$

Its interpretation is clear from the previous paragraphs.

By analogy the notations $H^u(w,\varepsilon)$, $h^*(w,\varepsilon)$, $w^{*\varepsilon}$, $H^u(W)$ are clear, but we need a definition of $H^*(W)$. For that let us introduce the lexicographic order, $\succcurlyeq$ , for sequences of real numbers, that is :

$$(y_0, y_1, \ldots) \succcurlyeq (y'_0, y'_1, y'_2, \ldots) \text{ is true} \Leftrightarrow (\text{if } y_n = y'_n, \forall n < m \text{ then } y_m \geq y'_m) \ \forall m \in \mathbb{N}. \tag{2.20}$$

We denote by $\vec{\min}$ the minimum for this order. Then we define $H^*$ by :

$$H^*_{.x}(W) = \vec{\min}\, H^u_{.x}(W) \tag{2.21}$$

(indeed $H^u_{.x}(W)$ is a sequence of real numbers).

We shall prove that $w^{*\varepsilon}$ admits an expansion in $\varepsilon$ denoted by $W^*(\varepsilon)$ which satisfies :

$$H^*(W^*) = 0 \tag{2.22}$$

The purpose of this paper is to prove this last result and to show that $W^*$ can be computed recursively. By this way we can design faster algorithm than the ones obtained by a direct solution of $h^*(w^{*\varepsilon},\varepsilon) = 0$.

## 3. – REVIEW OF MARKOV CHAINS

Let us recall some facts on Markov chains. We consider the Markov chain defined by $(T,\mathcal{X}, m, c, \lambda)$.

The matrix m defines a connexity in the state space $\mathcal{X}$, that is : $x \in \mathcal{X}$ and $x' \in \mathcal{X}$ are connected if there exists a non zero probability path between x and x'. Moreover if x' and x are also connected we say that x and x' are strongly-connected. The equivalence classes of the strongly-connexity relation defines a partition on $\mathcal{X}$. The connexity relation defines a partial order on these classes. The final classes for this order are the recurrent classes of the Markov chain. Their set is denoted by $\bar{\mathcal{X}}_r = \{\bar{x}_1, \bar{x}_2, \dots, \bar{x}_{|\bar{\mathcal{X}}_r|}\}$. The other states are called transient and their set is denoted by $\bar{x}_t$. Thus we have defined a partition $\bar{\mathcal{X}}$ of $\mathcal{X}$, $\bar{\mathcal{X}} = \bar{x}_t \cup \bar{\mathcal{X}}_r$. Let us consider the natural numerotation of the states of $\mathcal{X}$ after the grouping defined by the partition $\bar{\mathcal{X}}$. With this numeration the transition matrix has the following block structure :

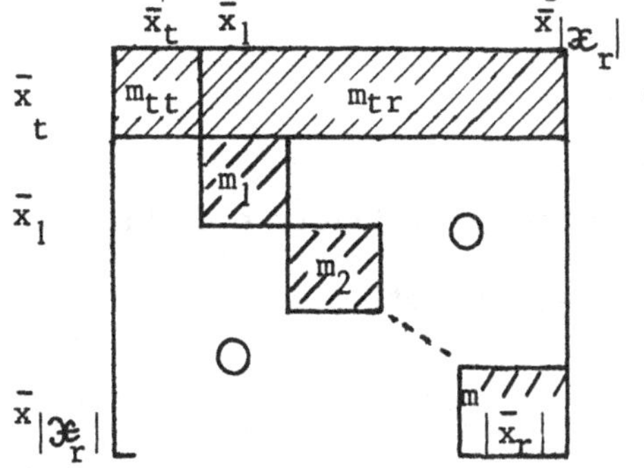


$m_{tr}$ has at least a non zero term by row

(3.1)

m admits the eigenvalue 1 because $\sum_{x' \in \mathcal{X}} m_{xx'} = 1$. This eigenvalue is semi-simple (the eigen-space associated to the eigenvalue 1 admits a base of eigenvectors). This can be proved easily by remarking that $|m|_{\infty,\infty} = 1$ which proves that $|m^n|_{\infty,\infty} = 1$, where $|m|_{\infty,\infty}$ denotes the norm of matrices seen as operators on $\mathbb{R}^{|\mathcal{X}|}$ with the sup norm. Now if the eigenvalue 1 was not semi-simple, in an appropriate basis m would have a Jordan block : $\begin{bmatrix} 1 & 1 & 0 \\ & 1 & 1 \\ 0 & & 1 \end{bmatrix}$ and we should have $|m^n|_{\infty,\infty} \xrightarrow[n\to\infty]{} \infty$ which is a contradiction. From this property we see that we have the decomposition :

$$\mathbb{R}^{\mathcal{X}} = \mathcal{N}(a) \oplus \mathcal{R}(a) \tag{3.2}$$

where :

$$a \equiv m - i \tag{3.3}$$

$\mathcal{N}(a)$ denotes the kernel of $\mathcal{R}(a)$, the range of the operator a.

To define the projector $a^0$ on $\mathcal{N}(a)$ parallel to $\mathcal{R}(a)$ we need to know a base of $\mathcal{N}(a)$ and $\mathcal{N}(a')$ where ' denotes the transposition.

The set $\{p_{\bar{x}.}, \bar{x} \in \bar{\mathcal{X}}_r\}$ of the extremal invariant probability measures of m defines a base of $\mathcal{N}(a')$. $p_{\bar{x}}$ has for support $\bar{x}$ and the restriction to $\bar{x}$ of $p_{\bar{x}.}$ denoted by $\bar{p}_{\bar{x}}$ satisfies :

$$\bar{p}_{\bar{x}} m_{\bar{x}} = \bar{p}_{\bar{x}}. \tag{3.4}$$

This result is clear from (3.1).

The set $\{q_{.\bar{x}}, \bar{x} \in \bar{\mathcal{X}}_r\}$ where $q_{x\bar{x}}$ denotes the probability starting from x to end in $\bar{x}$ defines a base of $\mathcal{N}(a)$. Indeed $q_{\bar{x}}$ satisfies :

$$q_{x\bar{x}} = \begin{cases} 0 & \text{if } x \notin \bar{x} \cup \bar{x}_t \\ 1 & \text{if } x \in \bar{x} \\ \bar{q}_{\bar{x}} & \text{if } x \in \bar{x}_t \end{cases} \tag{3.5}$$

with $\bar{q}_{\bar{x}}$ solution of the Dirichlet problem :

$$m_{tt}\, \bar{q}_{\bar{x}} = - m_{t\bar{x}}\, 1^{*} \tag{3.6}$$

From (3.5) and (3.6) it is clear that $q_{\bar{x}} \in \mathcal{N}(a)$, $(q_{\bar{x}},\ \bar{x} \in \bar{\mathfrak{X}}_r)$ are linearly independent and from (3.1) that they form a base of $\mathcal{N}(a)$.

If we see p as $(|\bar{\mathfrak{X}}|, |\mathfrak{X}|)$-matrix and q as a $(|\mathfrak{X}|, |\bar{\mathfrak{X}}|)$-matrix the projector on $\mathcal{N}(a)//\mathfrak{R}(a)$ is :

$$a^{0} = q\ p. \tag{3.7}$$

We have :

$$aa^{0} = a^{0}a = 0 \tag{3.8}$$

There exists a pseudo inverse $a^{+}$ of $-a$ which is the inverse of $-a$, restricted to $\mathfrak{R}$ (a), defined precisely by the relations:

$$\begin{cases} a^{+}a = a^{+}a = a^{0} - i \\ a^{+}a^{0} = a^{0}a^{+} = 0 \end{cases} \tag{3.9}$$

$\tau$ is a random time, that is a random variable on T, independent of the Markov chain $X_t$, of exponentiel probability law of parameter $\lambda$ that is :

$$P(\tau = t) = \frac{\lambda}{(1+\lambda)^{t+1}} \tag{3.10}$$

We have :

$$\mathbb{E}(\tau) = \frac{1}{\lambda} \tag{3.11}$$

On the new probability space $\tilde{\Omega} = \Omega \otimes T$ we have :

---

* 1 denotes the $|\bar{x}|$-vector : $1_x = 1,\ \forall x \in \bar{x}$.

$$\lambda\, w_x = \mathbb{E}\,[\sum_{t=0}^{+\infty} \frac{\lambda}{(1+\lambda)^{t+1}}\, c_{X(t,\omega)} \mid X(0,\omega) = x]$$
$$= \mathbb{E}[c_{X(\tau(\tilde\omega),\tilde\omega)} \mid X(0,\tilde\omega) = x] \tag{3.12}$$

The operator :

$$r_\lambda(a) : \mathbb{R}_c^{\mathfrak{X}} \to \mathbb{R}_w^{\mathfrak{X}} \tag{3.13}$$

is called the resolvent of a.

From (3.12) we see that $\lambda r_\lambda$ defines a transition matrix :

$$[\lambda r_\lambda(a)]_{xx'} = \mathbb{P}\,\{X[\tau(\tilde\omega),\tilde\omega] = x' \mid X(0,\tilde\omega) = x\},\ \forall x,x' \in \mathfrak{X} \tag{3.14}$$

which corresponds to the initial Markov seen at random time $\tau_1, \tau_2, \ldots, \tau_n$ with $\tau_{i+1} - \tau_i$ independent of $\tau_i$ and having the same probability law as $\tau$.

From the Jordan form of a and the previous discussion on the semi-simple nature of the eigenvalue 1 of m, we can show the ergodic theorem :

$$\lim_{\lambda\to 0} \lambda r_\lambda(a) = a^0. \tag{3.15}$$

## 4. - PERTURBED MARKOV CHAIN

We study the perturbed Markov chain $(T, \mathfrak{X}, \mathcal{E}, m(\varepsilon), c(\varepsilon), \lambda(\varepsilon))$, in the case $\lambda = \varepsilon^\ell \mu$, $v(c) = \ell$; that is, we study the transfer function $\varepsilon^\ell \mu(\varepsilon^\ell \mu + i - m(\varepsilon))^{-1}$ in $\varepsilon$. With the interpretation (3.14), this means that we look at the Markov chain on the time scale $\frac{t}{\varepsilon^\ell}$ ; for time scale interpretation in time domain see also Coderch-Sastry-Willsky-Castanon [2],[3]. We have seen in (2.11) that when the optimal conditional expected cost $w^\varepsilon$ admits an expansion, $W(\varepsilon)$, in $\varepsilon$ this expansion satisfies :

$$H(W) \equiv (M-I-\Lambda)W + C = 0 \tag{4.1}$$

(4.1) is an infinite set of linear equations. Conversely if a solution of

(4.1) exists with for example $(W_i, \; i \in \mathbb{N})$ bounded then $W(\varepsilon)$ converges, for $\varepsilon < 1$, and is a solution of :

$$h(w,\varepsilon) = 0 \tag{4.2}$$

Let us show now that (4.1) can be computed recursively.

For that we build the implicit realization of W :

$$\begin{cases} E \, y_{n+1} = F \, y_n - G \, C_{n+\ell+1} \\ W_{n+1} = H \, y_{n+1} \end{cases} \qquad y_{-1} = 0 \tag{4.3}$$

with :

$$a_0 = m_0 - i \tag{4.4}$$

$$E = \begin{bmatrix} a_0 & & & 0 \\ m_1 & \ddots & & \\ \vdots & \ddots & \ddots & \\ -\mu & \cdots & m_1 & a_0 \end{bmatrix} \quad (\ell+1) \text{ blocks} \tag{4.5}$$

$$F = \begin{bmatrix} 0 & a_0 & & 0 \\ \vdots & m_1 & \ddots & \\ \vdots & 0 & m_1 & a_0 \\ 0 & \cdots & \cdots & 0 \end{bmatrix} \quad (\ell+1) \text{ blocks} \tag{4.6}$$

$$G = \begin{bmatrix} 0 \\ \vdots \\ 0 \\ i \end{bmatrix} \quad (\ell+1) \text{ blocks} \tag{4.7}$$

$$H = \underbrace{[i \quad 0 \; \cdots \; 0]}_{(\ell+1)\text{blocks}} \tag{4.8}$$

Indeed if W is a solution of (4.1) :

$$y_n = (W_n, W_{n+1}, \dots, W_{n+\ell})$$

is a solution of (4.3).

Conversely if W is a solution of (4.3), by elimination of the variables y we see that W satisfies (4.1).

To prove the existence of a solution of (4.3), following Bernhard [1] we have to show that there exists $\mathcal{Z} \subset \mathbb{R}$ which satisfies :

$$F\mathcal{Z} \subset E\mathcal{Z}, \tag{4.9}$$

$$G \subset E\mathcal{Z}. \tag{4.10}$$

We can take $\mathcal{Z} = \mathbb{R}^{|\mathcal{X}|\times(\ell+1)}$. Indeed (4.9) is equivalent to finding a $z \in \mathbb{R}^{\mathcal{X}\times(\ell+1)}$ such that:

$$Ez = Fy \; ; \quad \forall y \in \mathbb{R}^{|\mathcal{X}|\times(\ell+1)}. \tag{4.11}$$

But by the change of variables $z'^k = z^k - y^{k+1}$, (4.11) becomes :

$$Ez' = Gc \text{ with } c = -\mu\, y^2 + a_1\, y^\ell \in \mathbb{R}^{\mathcal{X}} \tag{4.12}$$

which is a relation of (4.10) kind.

Delebecque [5] has proved that (4.10) has a solution. Let us show this result in two cases $\ell = 1$ and $\ell = 2$; then the general proof can be induced easily.

$\ell = 1$

We have to solve :

$$\begin{cases} a_0\, W_0 = 0 \\ (m_1-\mu) W_0 + a_0\, W_0 = -C_1. \end{cases} \tag{4.13}$$

(4.13) implies :

$$a_0^0 \, W_0 = W_0 \tag{4.14}$$

where $a_0^0$ denotes the projector on $\mathcal{N}(a_0) \;//\, \mathcal{R}(a_0)$.

Then left multiplying (4.13) by $a_0^0$ gives :

$$a_0^0(\mu-m_1)a_0^0 \, W_0 = a_0^0 \, C_1 \tag{4.15}$$

Using the factorization of $a_0^0 = qp$ where q and p are defined as in (3.4) and (3.5) for $m = m_0$ we have :

$$W_0 = q(\mu-p \; m_1 \; q)^{-1} \; p \; C_1 \qquad \forall \mu \notin \text{Spect.}(p \; m_1 \; q) \tag{4.16}$$

Thus :

$$W_1 = a_0^+ \; [(m_1-\mu)W_0 + C_1] + \bar{W}_1 \quad \forall \bar{W}_1 \; : \; \bar{W}_1 = a_0^0 \; \bar{W}_1 \tag{4.17}$$

where $a_0^+$ denotes the pseudo inverse of $-a_0$ defined in (3.9).

Then we have proved that $W_0$ is defined uniquely and $W_1$ up to an element of $\mathcal{N}(a_0)$.

From the stochastic interpretation :

$$W_0 = \lim_{\varepsilon\to 0} \mu \; \varepsilon(1 + \mu \; \varepsilon- m(\varepsilon))^{-1} \frac{C(\varepsilon)}{\mu \; \varepsilon} , \tag{4.18}$$

it follows that $p \; m_1 \; q$ is a generator of a Markov chain and thus that $(\mu - p \; m_1 \; q)^{-1}$ exists $\forall \mu > 0$.

<u>Example</u>

Consider the Markov chain :

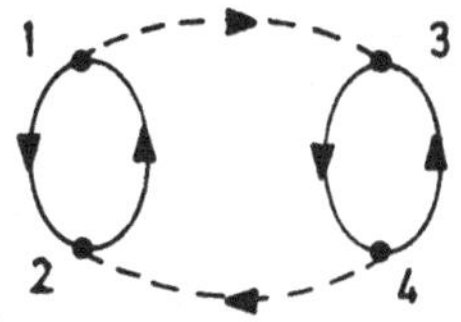

where the dotted lines corresponds to probabilities of order $\varepsilon$, the other lines to order-1 ones.

$m_0$ has the following block structure :

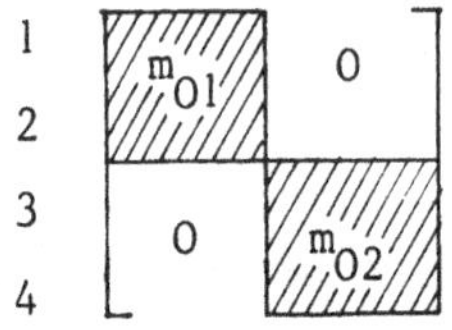


The dimension of $\mathcal{N}(a_0)$ is 2.

$$q = \begin{bmatrix} 1 & 0 \\ 1 & 0 \\ 0 & 1 \\ 0 & 1 \end{bmatrix} \tag{4.19}$$

$$p = \begin{bmatrix} \bar{p}_1^1 & \bar{p}_1^2 & 0 & 0 \\ 0 & 0 & \bar{p}_2^1 & \bar{p}_2^2 \end{bmatrix} \tag{4.20}$$

with :

$$\bar{p}_1 \, m_{01} = \bar{p}_1 \tag{4.21}$$

$$\bar{p}_2 \, m_{02} = \bar{p}_2 \tag{4.22}$$

The aggregated chain of generator $p_0 \, m_1 \, q_0$ is :

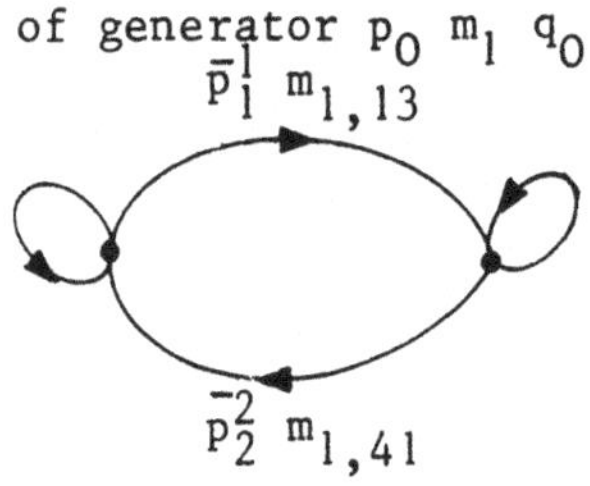


$\underline{\ell = 2}$

We have to solve :

$$\begin{cases} a_0 \, W_0 = 0 \\ m_1 \, W_0 + a_0 \, W_1 = 0 \\ -\mu \, W_0 + m_1 \, W_1 + a_0 \, W_2 = - C_2 \end{cases} \tag{4.23}$$

Left multiplying (4.23) by $a_0^0$ we have :

$$\left\{\begin{array}{l} a_0^0\, W_0 = W_0 \\ a_0^0\, m_1\, a_0^0\, W_0 = 0 \qquad W_1 = a_0^+\, m_1\, W_0 + a_0^0\, W_1 \\ a_0^0\, m_1\, W_1 - \mu\, a_0^0\, W_0 = -\, a_0^0\, C_2 \end{array}\right. \tag{4.24}$$

which gives using the formula of $W_1$ :

$$\left\{\begin{array}{l} a_0^0\, W_0 = W_0 \\ a_0^0\, m_1\, a_0^0\, W_0 = 0 \\ a_0^0\, m_1\, a_0^0\, W_1 + a_0^0\, m_1\, a_0^+\, m_1\, a_0^0\, W_0 - \mu\, a_0^0\, W_0 = -a_0^0\, C_2 \end{array}\right. \tag{4.25}$$

Using the notations :

$$\left\{\begin{array}{l} \bar{a}_0 = p\, m_1\, q \\ \bar{m}_1 = p\, m_1\, a_0^+\, m_1\, q \\ \bar{W}_0 = p\, W_0 \\ \bar{W}_1 = p\, W_1 \\ \bar{C}_1 = a_0^0\, C_2 \end{array}\right. \tag{4.26}$$

(4.25) becomes:

$$\left\{\begin{array}{l} \bar{a}_0\, \bar{W}_0 = 0 \\ (\bar{m}_1 - \mu)\bar{W}_0 + \bar{a}_0\, \bar{W}_1 = -\bar{C}_1 \end{array}\right. \tag{4.27}$$

(4.26) is a problem of kind $\ell = 1$. Thus using the factorization of $\bar{a}_0^0 = \bar{p}\, \bar{q}$ which exists because $\bar{a}_0$ is a generator of a Markov chain (see $\ell = 1$ case), we obtain :

$$W_0 = q\, \bar{q}\, (\mu - \bar{p}\, \bar{m}_1\, \bar{q})^{-1}\, \bar{q}\, q\, C_2 \quad \forall \mu \notin \text{Spect.}(\bar{p}\, \bar{m}_1\, \bar{q}) \tag{4.28}$$

Then from (4.27) and (4.24) we can compute $W_1$ and $W_2$. $W_1$ is defined up to an element of $q\mathcal{N}(\bar{a}_0)$, $W_2$ up to an element of $\mathcal{N}(a_0)$.

The stochastic interpretation of $W_0$ :

$$W_0 = \lim_{\varepsilon \to 0} \mu\varepsilon^2 \, (1+\varepsilon^2\mu - m(\varepsilon))^{-1} \frac{C(\varepsilon)}{\mu\,\varepsilon^2} \tag{4.29}$$

shows that $\bar{p}\,\bar{m}_1\,\bar{q}$ is a generator of a Markov chain. Thus we have the existence of (4.21), $\forall\mu > 0$. ∎

This procedure can be reiterated and gives the general case (Delebecque [5]). In this reference we find the relation of this method and the reduction process of Kato [9]. ∎

The reiteration of the reduction process finishes when the aggregate chain obtained has the same number of recurrent classes as the one of the initial chain $(m(\varepsilon))$. ∎

Bernhard [1] has proved that the solution of the implicit system is unique if $\mathcal{N}(E) \cap \mathcal{Z} = \emptyset$. The discussion of the two cases $\ell = 1$ and $\ell = 2$ shows that $\mathcal{N}(E) \cap \mathbb{R}^{|\mathcal{X}| \times (\ell+1)} \neq 0$ in the case $\ell = 1$ :

$$\mathcal{N}(E) \cap \mathbb{R}^{2|\mathcal{X}|} = \mathcal{N}\left(\begin{bmatrix} i & 0 \\ 0 & a_0 \end{bmatrix}\right), \tag{4.30}$$

In the case $\ell = 2$, on an appropriate basis :

$$\mathcal{N}(E) \cap \mathbb{R}^{3|\mathcal{X}|} = \mathcal{N}\left(\begin{bmatrix} i & & 0 \\ & i_0 & \\ 0 & \bar{a}_0 & \\ & & a_0 \end{bmatrix}\right) \tag{4.31}$$

where $i_0$ denotes the identity on $\mathcal{R}(a_0)$.

But we see that the non unicity part of the implicit system is unobservable in the output; indeed $\mathcal{N}(H) \supset \mathcal{N}(E)$. This property is true in the general situation because we can prove that $W_0$ is always defined uniquely.

We have proved the :

Theorem 1 : The solution $W^{\varepsilon}$ of :

$$h(W,\varepsilon):=(m(\varepsilon)-i-\lambda(\varepsilon))W+c(\varepsilon)=0, \tag{4.32}$$

admits an expansion $W(\varepsilon)$ which is the unique solution of :

$$H(W):=(M-I-\Lambda)W+C=0 \tag{4.33}$$

Moreover W can be computed recursively by solving the implicit system realization of (4.36) :

$$\begin{cases} Ey_{n+1}=Fy_n-GC_{n+\ell+1},\ y_{-1}=0, \\ W_{n+1}=Hy_{n+1}, \end{cases} \tag{4.34}$$

where E,F,G,H are defined in (4.5) to (4.8).

This implicit system has an output uniquely defined and it admits a strictly causal realization.

The first term of the expansion has the interpretation of the conditional expected cost of an aggregated Markov chain obtained by reiteration of an aggregation procedure which consists in aggregating the recurrent classes of the order 1 transition matrix, in one state, and computing the transition matrix of the new aggregate chain.

## 5. - REVIEW OF CONTROLLED MARKOV CHAINS

Given the controlled Markov chain n-tuple: $(T,\mathcal{X},\mathcal{U},m^u,c^u,\lambda)$. The optimal conditional expected $w^*$ cost is the unique solution in w of the dynamic programming equation :

$$h^*_x(w)\equiv\min_u\,[(m^u-1-\lambda)w+c^u]_x=0,\ \forall x\in\mathcal{X}. \tag{5.1}$$

This result can be proved using the Howard algorithm :

Step 1 : Given a policy $s \in \mathcal{U}^{\mathscr{X}}$, let us compute w, solving, in w, the linear equation :

$$h^{os}(w) = 0 \tag{5.2}$$

Step 2 : Given a conditional expected cost w, let us improve the policy by computing :

$$\min_u h_x^u(w) \tag{5.3}$$

We change s(x) only if $h_x^u(w) < 0$. Then we return to step 1.

By this way we generate a sequence :

$$((s^n, w^n) \ ; \ n \in \mathbb{N})$$

which converges after a finite number of steps. The sequence $(w^n, n \in \mathbb{N})$ is decreasing.

Indeed :

$$h^{os^n}(w^n) = 0 \tag{5.4}$$

$$h^{os^{n+1}}(w^{n+1}) = 0 \tag{5.5}$$

Then (4.4)-(4.5) gives :

$$(m^{os^{n+1}} - 1 - \lambda)(w^n - w^{n+1}) + h^{os^n}(w^n) - h^{os^{n+1}}(w^n) = 0 \tag{5.6}$$

But by (4.3) we have :

$$h^{os^n}(w^n) - h^{os^{n+1}}(w^n) \geq 0 \tag{5.7}$$

Then (5.6) and (5.7) proves that :

$$w_n - w_{n+1} \geq 0 \tag{5.8}$$

Indeed, (5.6) can be seen as a Kolmogorov equation in $(w_n - w_{n+1})$, with a positive instantaneous cost.

The existence and the uniqueness of a solution in w of (4.1) follows easily from this result.

## 6. - CONTROL OF PERTURBED MARKOV CHAINS

Given the perturbed controlled Markov chain n-tuple $(T, \mathcal{X}, \mathcal{U}, \mathcal{E}, m^u(\varepsilon), c^u(\varepsilon), \lambda(\varepsilon))$. The optimal cost is the unique solution in w of the dynamic programming equation :

$$h^*_x(w,\varepsilon) \equiv \min_u[(m^u(\varepsilon)-1-\lambda(\varepsilon))w + c^u(\varepsilon)]_x = 0, \forall x \in \mathcal{X} \quad (6.1)$$

We have the :

Theorem 2 : The solution of (6.1) denoted by $w^{*\varepsilon}$ admits an expansion in $\varepsilon$ denoted by $W^*(\varepsilon)$ which is the unique solution in W of the vectorial dynamic programming equation :

$$H^*_{.x}(W) \equiv \overrightarrow{\min_u}[(M^u-I-\Lambda)W + C^u]_{.x} = 0, \forall x \in \mathcal{X} \quad (6.2)$$

Let us remember that $\overrightarrow{\min}$ means the minimum for the lexicographic order on the sequence of real numbers.

The solution $W^*$ can be computed by the vectoriel Howard algorithm :

Step 1 : Given a policy $s \in \mathcal{U}^{\mathcal{X}}$, let us compute W using the results of part 4 :

$$\text{Hos } (w) = 0 \quad (6.3)$$

Step 2 : Given a conditional expected cost W, let us improve the policy by computing :

$$\overrightarrow{\text{Min}_u}\; H^u_x(W) \quad (6.4)$$

We change s(x) only if $H^u_{.x}(W) \neg 0$. Then we return to step 1.

By this way we generate a sequence :

$$((s^n, W^n) \; ; \; n \in \mathbb{N})$$

which converges after a finite number of steps. The sequence $(W^n,\ n \in \mathbb{N})$ is decreasing for the lexicographic order $\succ$.

This decreasing property can be proved easily using the corresponding proof in the unperturbed case parts, and the following equivalence :

$$h_x^u(W(\varepsilon),\varepsilon) \geq h_x^{u'}(W(\varepsilon),\varepsilon) \iff H_{.x}^u(W) \not\succ H_{.x}^{u'}(W). \tag{6.5}$$

From this property the theorem can be proved easily.

A priori it is not clear if we may restrict the minimization to finite part of the infinite sequence.

The following result shows that this is possible and gives an estimate on the length of the sequence part on which we have to apply the lexicographic order minimization.

<u>Theorem 3</u> : <u>The vectoriel minimization (in 6.4) may be applied on the $\eta = (d^0(C) + (v(\lambda) + 2)|\mathcal{X}|)$ first terms of the sequence only without changing the convergence to the solution of th. 2.</u>

<u>Proof</u> : Let us show that :

$$H_n^u(W) = H_n^{u'}(W),\ \forall i = d^0(c)+1,\dots,\eta \implies H_n^u(W) = H_n^{u'}(W), \forall \eta > d^0(c) \tag{6.6}$$

By theorem 1, W admits a strictly causal realization that is there exists $\tilde{E}$, $\tilde{F}$, $\tilde{G}$ such that :

$$\left\{\begin{aligned} z_{n+1} &= \tilde{E}\, z_n + \tilde{F}\, C_{n+\ell+1} \\ W_{n+1} &= \tilde{G}\, z_{n+1} \end{aligned}\right. \tag{6.7}$$

For that we have equal to zero the control corresponding to the non-unicity of the implicit system (4.34) because this non-unicity is not observable. By (4.34) we know that the order of the matrix E is smaller than $(v(\lambda)+1)|\mathfrak{X}|$. The entry $C_{n+\ell+1}$ is equal to zero for $n \geq d^0(c)-v(c)$.

We add $|\mathfrak{X}|$ new states to z, denoted by $\tilde{z}$ with :

$$\tilde{z}_{n+1} = W_n. \tag{6.8}$$

With the new state $\check{z} = (z,\tilde{z})$ the second part of (6.6) can be written :

$$([a_0^u-a_0^{u'}][\tilde{G},0] + [m_1^u-m_1^{u'}][0,i])\ \check{z}_n = 0 \tag{6.9}$$

(6.9) has the form :

$$J\ \check{z}_n = 0 \tag{6.10}$$

with J an observation matrix of the dynamical system of state $\check{z}_n$. It follows by the Cayley-Hamilton theorem that if (6.10) is true $\forall n : \eta \geq n > d^0(c)$ then (6.10) is true $\forall n > d^0(c)$. The theorem 3 is deduced easily from this result.

Remark :

The order of E in 4.34 is $(v(\lambda)+1)|\mathfrak{X}|$ but the order of $\tilde{E}$ is much smaller. It is certainly of order $|\mathfrak{X}|$. Moreover it is certainly not necessary to memorize completely $V_n$ in (6.8) to be able to compute it from $z_{n+1}$, thus the value of $\eta$ is much smaller than the value given in the theorem 3.

Stochastic interpretation of the first term of the expansion can be found in Delebecque - Quadrat [6], [7].

REFERENCES

[1] P. BERNHARD. Sur les systèmes dynamiques linéaires implicites singuliers, à paraître SIAM J. on control and optimization et rapport INRIA 69, 1981.

[2] M. CODERCH, A.S. WILLSKY, S.S. SASTRY, D.A. CASTANON. Hierarchical aggregation of linear systems with multiple time scales, MIT Report LIDS-P-1187, mars 1982.

[3] M. CODERCH, A.S. WILLSKY, S.S. SASTRY. Hierarchical aggregation of singularly perturbed finite state Markov chains submitted to stochastics.

[4] P.J. COURTOIS. Decomposability, ACM Monograph Series, Academic Press, 1977.

[5] F. DELEBECQUE. A reduction process for perturbed Markov chains, à paraître SIAM J. of applied math. to appear.

[6] F. DELEBECQUE, J.P. QUADRAT. Optimal control of Markov chains admitting strong and weak interactions, Automatica, Vol. 17, n° 2, pp. 281-296, 1981.

[7] F. DELEBECQUE, J.P. QUADRAT. The optimal cost expansion of finite controls finite states Markov chains with weak and stong interactions. Analysis and optimization of systems, Lecture Notes an control and Inf. Science 28 Springer Verlag, 1980.

[8] A.A. PERVOZVANSKII, A.V. GAITSGORI. Decomposition aggregation and approximate optimization en Russe, Nauka, Moscou, 1979.

[9] T. KATO. Perturbation theory for linear operator, Springer Verlag, 1976.

[10] B.L. MILLER, A.F. VEINOTT. Discrete dynamic programming with small interest rate. An. math. stat. 40, 1969, pp. 366-370.

[11] R. PHILIPS, P.KOKOTOVIC . A singular perturbation approach to modelling and control of Markov chains IEEE A.C. Bellman issue, 1981.

[12] H. SIMON, A. ANDO. Aggregation of variables in dynamic systems, Econometrica, 29, 111-139, 1961 .

[13] J. KEMENY, L. SNELL. Finite Markov chains, Van Nostrand, 1960.

[14] O. MURON. Evaluation de politiques de maintenance pour un système complexe, RIRO, vol. 14, n° 3, pp. 265-282, 1980.

[15] S.L. CAMPBELL, C.D. MEYER Jr. Generalized inverses of linear transformations. Pitman, London, 1979.

# SINGULAR PERTURBATIONS IN STABLE FEEDBACK CONTROL OF DISTRIBUTED PARAMETER SYSTEMS

Mark J. Balas
Associate Professor
Electrical, Computer, and Systems Engineering Department
Rensselaer Polytechnic Institute
Troy, New York 12181 U.S.A.

ABSTRACT

In this lecture we use a singular perturbation formulation of linear time-invariant distributed parameter systems to develop a method to design finite-dimensional feedback compensators of any fixed order which will stabilize the infinite-dimensional distributed parameter system. The synthesis conditions are given entirely in terms of a finite-dimensional reduced-order model; the stability results depend on an infinite-dimensional version of the Klimushchev-Krasovskii lemma also presented here. This lecture summarizes our work on singular perturbations for stable distributed parameter system control in [9]-[10] and[24].

## 1.0 INTRODUCTION

Many engineering systems exhibit a distributed parameter nature and, in order to be accurately modeled, they must be described by partial differential equations. Examples of such distributed parameter systems (DPS) include heat diffusion and chemical processes, wave propagation, and mechanically flexible structures. Various aspects of the control of DPS have been considered in, for example, [1]-[5]; our experience in DPS has been shaped by applications in large aerospace structures [6].

The state spaces for DPS have infinite dimension; so, at best, reduced-order models must be used in controller synthesis. However, the closed-loop stability of the infinite-dimensional DPS with a finite-dimensional feedback controller becomes a fundamental issue. The synthesis of finite-dimensional controllers for DPS and the analysis of their closed-loop stability by singular and regular perturbation techniques have been our main areas of emphasis [7]; this theory has been developed with flexible structures and other highly oscillatory DPS applications in mind.

Even in large-scale, lumped parameter systems, such as electric power distribution networks, it is necessary to perform model reduction and reduced-order controller synthesis and to analyze closed-loop stability. The use of asymptotic methods, especially singular perturbations, has been very successful in this regard (e.g., [8]). We have extended certain of these singular perturbations methods for DPS to provide estimates of stability in an infinite-dimensional setting [9]-[10] and [24] and applied them to mechanically flexible structures [11]. In this lecture we will describe these singular perturbation results and use them to synthesize general finite-dimensional compensators for linear DPS and show that they stabilize the infinite-dimensional closed-loop system.

A large number of the DPS applications have a singular parameter $\varepsilon$ representing time or frequency scaling or other small effects. Here we will deal with linear DPS having a <u>singular perturbation formulation</u>:

$$\begin{cases} E(\varepsilon)\,\dfrac{\partial v(t)}{\partial t} = A\,v(t) + Bf(t);\; v(o) = v_o \\ y(t) = C\,v(t) \end{cases} \tag{1.1}$$

where the state v(t) is in an infinite-dimensional Hilbert space H with inner product denoted by $(\cdot,\cdot)$ and corresponding norm $||\cdot||$. The operator A is a closed, linear, unbounded differential operator with domain D(A) dense in H, and A generates a $C_o$-semigroup of bounded operators U(t) on H. The operators B & C have finite ranks M & P, respectively, and f(t), y(t) represent the inputs from M actuators and the outputs from P sensors, respectively. Thus,

$$B\,f(t) = \sum_{i=1}^{M} b_i\,f_i(t) \tag{1.2}$$

and

$$y(t) = [y_1(t),\ldots,y_P(t)]^T \text{ where}$$

$$y_j(t) = (c_j, v(t)) \qquad 1 \leq j \leq P \tag{1.3}$$

with $b_i$ and $c_j$ in H. The linear operator $E(\varepsilon):H\to H$ is a continuous function of the singular parameter $\varepsilon \geq 0$. It has a bounded inverse when $\varepsilon>0$ but E(o) is singular; hence, (1.1) is called a singular perturbation formulation, as opposed to the usual DPS formulation where $E(\varepsilon)=I$ (see [7]).

Feedback control for such a DPS must be accomplished with finite-dimensional, discrete-time controllers of the form:

$$\begin{cases} f(k)=L_{11}y(k)L_{12}z(k) \\ z(k+1)=L_{21}y(k)+L_{22}z(k) \end{cases} \tag{1.4}$$

where z belongs to $R^{\alpha}$. Such controllers can be implemented with on-line digital computers whose memory access time is related to the controller dimension $\alpha$. Controller synthesis is based on reduced-order models of the DPS which can be obtained by assuming $\varepsilon=0$. However, the stability of such a feedback controller in closed-loop with the actual DPS, where $\varepsilon$ is small and positive but not zero, is in question. For convenience, we shall obtain our results for the continuous-time version of (1.4); see (3.1). This will focus attention on the essential problem of stable control of an infinite-dimensional system by a finite-dimensional controller; the implementation of a continuous-time control law with a digital computer is certainly a nontrivial issue but one of less theoretical magnitude.

Unlike the situation in finite-dimensional system theory, there are many types of DPS stability depending on the various types of convergence in infinite-dimensional spaces. However, exponential stability is the one of primary interest in engineering systems. A DPS of the form:

$$\begin{cases} \frac{\partial v}{\partial t} = A_c v \\ v(o) = v_o \end{cases} \tag{1.5}$$

is exponentially stable if $A_c$ generates a $C_o$-semigroup $U_c(t)$ with the growth property:

$$||U_c(t)|| \leq K_c e^{-\sigma_c t}, t \geq 0 \tag{1.6}$$

where $K_c$ and $\sigma_c$ are constants with $K_c \geq 1$ and $\sigma_c > 0$. This means that all solutions of (1.5) converge exponentially to zero with a rate $\sigma_c$.

Model reduction for DPS using the singular perturbations formulation (1.1) is the subject of Sec. 2.0. In Sec. 3.0, the synthesis of finite-dimensional controller-compensators is addressed. An infinite-dimensional version of the important Klimushchev-Krasovskii lemma is presented in Sec. 4.0. Our main results on the closed-loop stability are based on this lemma; they are given in Sec. 5.0 for the special case of reducing subspaces first and then extended to the general case of non-reducing subspaces. Our conclusions and recommendations appear in Sec. 6.0.

An expanded version of this lecture appears as [24]. The proofs of all results are omitted here but appear in [24]; furthermore, that reference discusses the situation where (A,B,C) in (1.1) depend on $\varepsilon$.

## 2.0 REDUCED-ORDER MODELING OF DPS: A SINGULAR PERTURBATIONS FORMULATION

Since the state space H of the DPS in (1.1) is infinite-dimensional, we must obtain a reduced-order model (ROM) upon which to base the finite-dimensional controller design. In general, this is done by selecting a finite-dimensional subspace $H_N$ (with dim $H_N=N<\infty$) contained in D(A). This subspace $H_N$ is the ROM subspace; its complement $H_R$ is the Residual subspace, and together they decompose the state space:

$$H=H_N\oplus H_R \tag{2.1}$$

We define the (not necessarily orthogonal) projection operators $P_N$ and $P_R$ and let $v_N=P_Nv$ and $v_R=P_Rv$. These decompose the state $v=v_N+v_R$ and the DPS (1.1).

We define a DPS singular perturbations model reduction as a pair $(H_N,H_R)$ of subspaces of H for which the following are satisfied:

(a) $H_N\subseteq D(A)$ and dim $H_N=N<\infty$;

(b) $H_N$ and $H_R$ decompose the state space (i.e., (2.1) is satisfied);

(c) the projections $(P_N,P_R)$ intertwine with $E(\varepsilon)$ in the following way:

$$\begin{cases} P_N E(\varepsilon) = P_N \\ P_R E(\varepsilon) = \varepsilon P_R \end{cases} \tag{2.2}$$

These assumptions yield the decomposed DPS (1.1) in the following form:

$$\frac{\partial v_N}{\partial t} = A_N v_N + A_{NR} v_R + B_N f \tag{2.3a}$$

$$\varepsilon \frac{\partial v_R}{\partial t} = A_{RN} v_N + A_R v_R + B_R f \tag{2.3b}$$

$$y = C_N v_N + C_R v_R \tag{2.3c}$$

where $A_N = P_N A\ P_N$, $A_{NR} = P_N A\ P_R$, $B_N = P_N B$, $A_{RN} = P_R A\ P_N$, $A_R = P_R A\ P_R$, $B_R = P_R B$, $C_N = C\ P_N$, and $C_R = C\ P_R$. All operators except $A_R$ are bounded (in fact, they have finite rank and, hence, are compact). The terms $A_{NR} v_R$ and $A_{RN} v_N$ are called <u>modeling error</u>, and terms $B_R f$ and $C_R v_R$ are called <u>control and observation spillover</u>, respectively; these terms represent the interconnections through which the feedback controller can affect the residual subsystem.

In the special case where $(H_N, H_R)$ are <u>reducing</u> (or modal) <u>subspaces</u>, we have [7]:

$$A_{NR} = 0 \text{ and } A_{RN} = 0 \tag{2.4}$$

Such reducing subspaces exist when a finite number of eignevalues of A can be separated by a closed curve from the rest of the spectrum of A; see [12] Theo. 6.17 p 178. Of course, $H_N$ is the corresponding eigensubspace for this group of eigenvalues; however, in order to calculate the parameters of the decomposition (2.3), the exact eigenfunctions (modes) must be known. In practice, this is not always possible; consequently, approximate modes are used and this leads to the nonzero modeling error terms. Two other special cases are (1) no control spillover when all $b_i$ belong to $H_N$ and (2)—no observation spillover when all $c_j$ belong to $H_N^{\perp}$; these situations would be very difficult to achieve with practical actuators and sensors.

The reduced-order model (ROM) of the DPS in this formulation is obtained by setting $\varepsilon=0$ in (2.3):

$$\begin{cases} \frac{\partial v_N}{\partial t} = \tilde{A}_N v_N + \tilde{B}_N f \\ y = \tilde{C}_N v_N + \tilde{D}_N f \end{cases} \tag{2.5}$$

where $\tilde{A}_N \equiv A_N - A_{NR} A_R^{-1} A_{RN}$, $B_N \equiv B_N - A_{NR} A_R^{-1} B_R$, $C_N \equiv C_N - C_R A_R^{-1} A_{RN}$, and $\tilde{D}_N \equiv -C_R A_R^{-1} B_R$. In general, this singular perturbations ROM is different from the usual ROM $(A_N, B_N, C_N)$ obtained by ignoring the residual subsystem [7] because of the static correction terms. Even when reducing subspaces are used and $\tilde{A}_N = A_N$, $\tilde{B}_N = B_N$, $\tilde{C}_N = C_N$, the feedthrough term $\tilde{D}_N$ is present in (2.5). All operators in the ROM (2.5) have finite rank and can be identified with their matrices in a basis of the subspace $H_N$; these matrices are useful for the specific controller synthesis but are not important for our purposes here. The ROM is completely determined by the choice of subspaces $H_N$ and $H_R$ in the singular perturbations model reduction process. Since there may be more than one singular perturbations formulation (1.1), there may be several ROM's; this can be used to the control designer's advantage.

Henceforth we will make <u>two basic assumptions</u> about the model reduction process:

(a) The ROM $(\tilde{A}_N, \tilde{B}_N, \tilde{C}_N, \tilde{D}_N)$ is controllable and observable;

(b) the residual subsystem is exponentially-stable, i.e. $A_R \equiv P_R A\, P_R$ generates a $C_o$-semigroup $U_R(t)$ with the growth property:

$$||U_R(t)|| \leq K_R e^{-\sigma_R t}, \quad t \geq 0 \tag{2.6}$$

where $K_R \geq 1$ and $\sigma_R > 0$. In some applications, $\sigma_R$ may be quite small, as it is in the case of large-space structures [6]. The first assumption is easily verified by the usual finite-dimensional system tests (e.g., see [13]). The second assumption is much more critical since it deals with the infinite dimensional residual subsystem. In practical terms, it says that we never want to disregard unstable parts of the system during model reduction; in theoretical terms, it is sometimes difficult to verify and we present two basic tests: Hille-Yosida and dissipativity. From the well-known Hille-Yosida Theorem (e.g., see [14]), $A_R$ generates the $C_o$-semigroup $U_R(t)$ satisfying (2.6) if and only if

$$||R(\lambda, A_R)^n|| \leqslant \frac{K_R}{(\lambda+\sigma_R)^n}; \quad n=1,2,\ldots \tag{2.7}$$

for all real $\lambda > -\sigma_R$ in the resolvent set $\rho(A_R)$ of $A_R$. The operator $R(\lambda, A_R) = (\lambda I - A_R)^{-1}$ is the resolvent operator for $A_R$ and by definition it is a bounded linear operator on H for each $\lambda$ in $\rho(A_R)$; the spectrum of $A_R$ is defined to be the complement of $\rho(A_R)$ in the complex plane. In the special situation where $A_R$ is dissipative, i.e.

$$\begin{cases} (A_R v, v) + \sigma_R (v, v) \leq 0 \\ (A_R^* v, v) + \sigma_R (v, v) \leq 0 \end{cases} \tag{2.8}$$

for all v in $D(A_R)$ or $D(A_R^*)$ where $A_R^*$ is the adjoint operator for $A_R$, the constant $K_R = 1$ in (2.6); this is a direct consequence of Theo. 3.2 in [15]. When it holds, the dissipativity condition (2.8) is often easier to verify than (2.7) especially when $A_R$ is self-adjoint (i.e., $A_R^* = A_R$); however, it may happen that $A_R$ is not dissipative and yet (2.6), or equivalently (2.7), holds.

## 3.0 FINITE-DIMENSIONAL CONTROLLER-COMPENSATORS FOR DPS

The form of the finite-dimensional feedback controller-compensators used here will be the following:

$$\begin{cases} f = L_{11}y+L_{12}z \\ \dot{z} = L_{21}y+L_{22}z \end{cases} \tag{3.1}$$

where the compensator states z has dim $z = \alpha \leq N < \infty$ and $L_{11}, L_{12}, L_{21}, L_{22}$ are matrices of appropriate sizes. We say the compensator is output feedback when $\alpha=0$. The order of the compensator is assumed to be fixed at some acceptable value which reflects the available capacity of the on-line computer being used to implement (3.1). In [10], the compensator order was $\alpha=N$.

The compensator design is synthesized as though the ROM (2.5) were the full DPS (1.1), i.e., as though $\varepsilon=0$. Let

$$F_N = \bar{A}_N+\bar{B}_N\bar{L}\,\bar{C}_N \tag{3.2}$$

where

$$\bar{A}_N = \begin{bmatrix} \tilde{A}_N & 0 \\ 0 & 0 \end{bmatrix}, \quad \bar{B}_N = \begin{bmatrix} \tilde{B}_N & 0 \\ 0 & I \end{bmatrix}, \quad \bar{C}_N = \begin{bmatrix} \tilde{C}_N & 0 \\ 0 & I \end{bmatrix}, \text{ and}$$

$$\bar{L} = \begin{bmatrix} \bar{L}_{11} & \bar{L}_{12} \\ \bar{L}_{21} & \bar{L}_{22} \end{bmatrix}$$

with

$$\begin{cases} \bar{L}_{11} = (I_M-L_{11}\tilde{D}_N)^{-1}L_{11} \\ \bar{L}_{12} = (I_M-L_{11}\tilde{D}_N)^{-1}L_{12} \\ \bar{L}_{21} = L_{21}(I_P+\tilde{D}_N\bar{L}_{11}) \\ \bar{L}_{22} = L_{22}+L_{21}\tilde{D}_N\bar{L}_{12} \end{cases} \tag{3.3}$$

Let the composite state $q_N = \begin{bmatrix} v_N \\ z \end{bmatrix}$ in $H_N x R^{\alpha}$; from (2.5) and (3.1), this satisfies (for $\varepsilon=0$):

$$\frac{\partial q_N}{\partial t} = F_N q_N \tag{3.4}$$

The following theorem gives the conditions under which a stable design can be synthesized:

Theorem 1: If

$$M+P+\alpha \geq N+1 \tag{3.5}$$

then $\bar{L}$ may be chosen so that $F_N$ given by (3.2) has almost any desired pole locations in the complex plane, i.e., the poles of $F_N$ are arbitrarily close to the desired ones.

Since $(\tilde{A}_N, \tilde{B}_N, \tilde{C}_N, \tilde{D}_N)$ is assumed controllable and observable, it is straightforward to see that the same is true of $(\bar{A}_N, \bar{B}_N, \bar{C}_N, \bar{D}_N)$. Now the proof follows easily from the finite-dimensional results of [16] or [17] because, when $\varepsilon=0$, the system is finite-dimensional. The operator $F_N$ may be identified with its matrix in some basis for $H_N x R^{\alpha}$ and the gains $\bar{L}$ chosen so that the eigenvalues of $F_N$ are located arbitrarily close to any desired values in the complex plane. Then the actual compensator gains $L_{ij}$ can be obtained from

$$\begin{cases} L_{11} = \bar{L}_{11}(I_P + \tilde{D}_N \bar{L}_{11})^{-1} \\ L_{12} = (I_M - L_{11}\tilde{D}_N)\,\bar{L}_{12} \\ L_{21} = \bar{L}_{21}(I_P - \tilde{D}_N \bar{L}_{11})^{-1} \\ L_{22} = \bar{L}_{22} - L_{21}\tilde{D}_N \bar{L}_{12} \end{cases}$$

These expressions require that $I_P+\tilde{D}_N\bar{L}_{11}$ be nonsingular; correspondingly, (3.3) requires that $I_M-L_{11}\tilde{D}_N$ be nonsingular. Since $\tilde{D}_N=-C_R A_R^{-1} B_R$, both will be nonsingular when the spillover terms are sufficiently small.

Consequently, the synthesis of the controller-compensator (3.1) is a finite-dimensional design based on "extended" output feedback stabilization of the ROM (2.5). The inequality (3.5) indicates the basic <u>trade-off</u> in this design: total number of control devices (M+P) vs. on-line computer capacity ($\alpha$). The total of these must exceed the ROM dimension (N) in order for the compensator (3.1) to achieve any desired level of stability. Of course, the compensator and the ROM ($\varepsilon=0$) produce a stable (finite-dimensional) closed-loop system; however, the stability of this same compensator in closed-loop with the actual DPS ($\varepsilon>0$) remains in question. This is the subject of the next two sections.

## 4.0 AN INFINITE-DIMENSIONAL VERSION OF THE KLIMUSHCHEV-KRASOVSKII LEMMA

In finite-dimensional spaces, the stability of closed-loop singularly perturbed systems is usually analyzed with the aid of the Klimushchev-Krasovskii (K-K) Lemma; see [8] and [19]-[20]. This lemma gives conditions under which linear singularly perturbed systems are uniformly asymptotically stable for small enough $\varepsilon$. More recently upper bounds have been calculated for the acceptable size of $\varepsilon$; see [21]-[22]. Unfortunately, none of the proofs of these results can be easily extended to infinite-dimensional spaces. Consequently, in this section we will give an infinite-dimensional version of the K-K Lemma; our result will include an upper bound on the acceptable size of $\varepsilon$.

Let $H_1$ and $H_2$ be (possibly infinite-dimensional) Hilbert spaces.

Consider the following singularly-perturbed closed-loop system:

$$\begin{cases} \dfrac{\partial\omega_1}{\partial t} = A_{11}\,\omega_1 + A_{12}\omega_2 \\ \varepsilon\dfrac{\partial\omega_2}{\partial t} = A_{21}\omega_1 + A_{22}\omega_2 \end{cases} \tag{4.1}$$

where $\omega_i$ is in $H_i$. We will assume the linear operators $A_{11}:H_1\to H_1$, $A_{12}:H_2\to H_1$, and $A_{21}:H_1\to H_2$ are all bounded in the following way:

$$\begin{cases} ||A_{11}|| \le M_1 < \infty \\ ||A_{12}|| \le M_2 < \infty \\ ||A_{21}|| \le M_3 < \infty \end{cases} \tag{4.2}$$

Furthermore, assume the (possibly unbounded) linear operator $A_{22}:D(A_{22}) \subseteq H_2\to H_2$ generates a $C_0$-semigroup $U_2(t)$ on $H_2$ with the growth property:

$$||U_2(t)|| \le K_2\, e^{-\sigma_2 t};\quad t \ge 0 \tag{4.3}$$

where $K_2 \ge 1$ and $\sigma_2 > 0$, i.e. $A_{22}$ is exponentially stable. Consequently, although $A_{22}$ may be unbounded, $A_{22}^{-1}$ and $A_{22}^{-1}A_{21}$ are bounded and we define these bounds:

$$\begin{cases} ||A_{22}^{-1}|| \le M_4 = \dfrac{K_2}{\sigma_2} < \infty \\ ||A_{22}^{-1}A_{21}|| \le M_5 \le M_4\, M_3 < \infty \end{cases} \tag{4.4}$$

Note that the first of these two bounds follows from the Hille-Yosida Theorem (see Sec. 2.0) since zero is in the resolvent set of $A_{22}$.

The reduced-system for (4.1) is obtained when $\varepsilon=0$:

$$\frac{\partial\omega_1}{\partial t} = A_1\omega_1 \tag{4.5}$$

where $A_1:H_1\to H_1$ is the bounded linear operator defined by:

$$A_1 = A_{11} - A_{12}A_{22}^{-1}A_{21} \tag{4.6}$$

with upper bound given by:

$$||A_1|| \leq M_6 = M_1 + M_2 M_5 \tag{4.7}$$

Since $A_1$ is bounded, it generates a $C_o$-semigroup $U_1(t)$ on $H_1$. If $A_1$ is exponentially stable (i.e., the spectrum of $A_1$ is contained in the open left half of the complex plane); then

$$||U_1(t)|| \leq K_1 e^{-\sigma_1 t}, \quad t \geq 0 \tag{4.8}$$

where $K_1 \geq 1$ and $\sigma_1 > 0$ (with $\sigma_1$ determined by the real part of the left-most point in the spectrum of $A_1$).

Now we can state a version of the K-K Lemma For Infinite-Dimensional Spaces:

Theorem 2: Let $H_c = H_1 \times H_2$ be the Hilbert space with norm:

$$||\omega|| = (||\omega_1||^2 + ||\omega_2||^2)^{1/2} \tag{4.9}$$

If the following conditions hold:

(a) $A_{11}$, $A_{12}$, $A_{21}$ are bounded linear operators,

(b) $A_{22}$ is exponentially stable, i.e. (4.3) holds,

(c) $A_1 = A_{11} - A_{12}A_{22}^{-1}A_{21}$ is exponentially stable, i.e. (4.8) holds,

then there exists $\varepsilon_o > 0$ such that, for each fixed $0 < \varepsilon \leq \varepsilon_o$, the linear operator

$$A_c^{(\varepsilon)} = \begin{bmatrix} A_{11} & A_{12} \\ \frac{A_{21}}{\varepsilon} & \frac{A_{22}}{\varepsilon} \end{bmatrix} \tag{4.10}$$

generates a $C_o$-semigroup $U_c(t,\varepsilon)$ which is exponentially stable, i.e.

$$||U_c(t,\varepsilon)|| \leq K_c e^{-\sigma_c t}, \quad t \geq 0 \tag{4.11}$$

where $K_c \geq 1$, $\sigma_c > 0$, and these constants are independent of $\varepsilon$. Furthermore, an upper bound for $\varepsilon_o$ is given by

$$\varepsilon_o < \frac{\sigma_2}{(r+M_5)K_2M_2(rM_2+M_6)} \cdot \min\left(\frac{\sigma_1}{K_1}, rM_2\right) \tag{4.12a}$$

or equivalently, with $0 < \delta < 1$,

$$\varepsilon_o < \frac{\sigma_2}{(1+\delta)M_5M_2K_2(\delta M_5M_2+M_6)} \cdot \min\left(\frac{\sigma_1}{K_1}, \delta M_5M_2\right) \tag{4.12b}$$

where $M_1$ through $M_6$ are given in (4.2) and (4.4), $K_1$ and $\sigma_1$ in (4.8) and $K_2$ and $\sigma_2$ in (4.3), the constant $0<r<M_5$ is defined later in Lemma 1, and the constant $K_c$ and $\sigma_c$ are given by:

$$\sigma_c = \min(\hat{\sigma}_1, \hat{\sigma}_2/\varepsilon_o) \tag{4.13}$$

$$K_c = K_1K_2(1+\alpha+\alpha^2)^{1/2}(1+\mu+\mu^2) \tag{4.14}$$

where

$$\mu = r+M_5$$

$$\hat{\sigma}_1 = \sigma_1-\varepsilon_oK_1M_2M_4(r+M_5)(rM_2+M_6) > 0 \tag{4.15}$$

$$\hat{\sigma}_2 = \sigma_2-\varepsilon_oK_2M_2(r+M_5) > 0 \tag{4.16}$$

$$\alpha = \frac{M_2\varepsilon_o}{|\varepsilon_o\hat{\sigma}_1-\hat{\sigma}_2|} \tag{4.17}$$

This theorem says that, if the reduced system ($\varepsilon=0$) given by (4.5) is exponentially stable (and if the subsystem $A_{22}$ is exponentially stable, also), then the full system (4.1) is exponentially stable for small $\varepsilon$. In fact, an upper bound for the size of $\varepsilon$ is given in (4.12). Although Theo. 2 is valid in finite-dimensional spaces, the upper bound in (4.12) is different from those given in [21]-[22] for finite-dimensional spaces. The proof of Theo. 2 requires the following lemma:

Lemma 1: There exists $\varepsilon_1>0$ such that, for any fixed $0<\varepsilon\leq\varepsilon_1$, the non-linear mapping $h:\Omega_r\rightarrow\mathcal{L}$ defined by

$$h(L) = H_{22}^{-1}H_{21}+H_{22}^{-1}L(H_{11}-H_{12}L) \tag{4.18}$$

where $\mathcal{L}$ is the Banach space of bounded linear operators $L:H_1\rightarrow H_2$ and $\Omega_r = \{L\varepsilon\mathcal{L}|\ ||L-H_{22}^{-1}H_{21}|| \leq r\}$ for any constant r satisfying:

$$0 < r < M_5 \tag{4.19}$$

has a unique fixed point $L^*=L^*(\varepsilon,r)$ in $\Omega_r$. Furthermore,

$$||L^*|| \leq r + M_5 \tag{4.20}$$

$$\varepsilon_1 = \frac{\gamma}{2M_4(rM_2+M_6)} \tag{4.21}$$

$$\gamma = \frac{2r}{r+M_5} < 1 \tag{4.22}$$

and the algorithm:

$$L_{k+1} = h(L_k);\ k=0,1,2... \tag{4.23}$$

converges to $L^*$ from any $L_0$ in $\Omega_r$ (in particular, $L_0=A_{22}^{-1}A_{21}$) with a convergence rate given by:

$$||L_k-L^*|| \leq \frac{\gamma^k}{1-\gamma}||L_1-L_0|| \tag{4.24}$$

where $\gamma$ is given in (4.22).

The proof of Lemma 1 uses a local contraction mapping argument (see [14] p 22 or [18] p 78) and is given in [24]. The proof of Theo. 2 appears in [24] also.

Note that $\gamma$ in (4.22) depends on r, which is bounded above by $M_5$, and the upper bound on $\varepsilon_0$ in (4.12) depends on r, also. If we take a value of r which is close to $M_5$, then we are making the upper bound on $\varepsilon_0$ larger (because the denominator of (4.12) increases as $r^2$ but the numerator increases at most as r); however, $\gamma$ will be close to one. This will cause the algorithm (4.23), which finds $L^*$, to converge more slowly, as (4.24) reveals. Hence, the actual calculation of $L^*$ would be difficult; however, we only need to know $L^*$ exists (which is guaranteed by Lemma 1) in order to assess the stability of (4.1). Consequently, we should choose the value of $\delta$ close to one in order to yield the largest upper bound (4.12) on $\varepsilon_0$.

## 5.0 CLOSED-LOOP STABILITY OF THE DPS WITH A FINITE-DIMENSIONAL CONTROLLER-COMPENSATOR

In Sec. 3.0, a general method for synthesizing finite-dimensional controller-compensators was developed; this synthesis is based on stabilizing the reduced-system (2.5) which occurs when $\varepsilon=0$. The most crucial question is whether such a finite-dimensional controller (3.1) can stabilize the infinite-dimensional DPS (1.1), or equivalently (2.3), when $\varepsilon>0$. The answer, which we will establish in this section, is that it will do so <u>when</u> $\varepsilon$ <u>is sufficiently small</u>; a bound on the acceptable size of $\varepsilon$ will be obtained from Theo. 2. To simplify this analysis, the results will be presented first for the special case of reducing subspaces (2.4), i.e. (2.3) will have the special form:

$$\begin{cases} \dfrac{\partial v_N}{\partial t} = A_N v_N + B_N f & (5.1a) \\ \varepsilon \dfrac{\partial v_R}{\partial t} = A_R v_R + B_R f & (5.1b) \\ y = C_N v_N + C_R v_R & (5.1c) \end{cases}$$

Unfortunately, this special case, although theoretically interesting, has less practical interest; therefore, we will follow it with stability results for the general, or nonreducing subspace, case.

We will need the following lemma on linear operators:

Lemma 2: Let $H_1$ and $H_2$ be normed linear spaces, and let $W:H_1 \to H_2$, $Q:H_2 \to H_1$ be bounded linear operators. If the inverses $(I_1+QW)^{-1}$ and $(I_2+WQ)^{-1}$ both exist, where $I_i$ is the identity operator on $H_i$, then the following operator identity holds:

$$(I_1+QW)^{-1} = I_1 - Q(WQ+I_2)^{-1}W \tag{5.2}$$

The proof of this is in [24].

## 5.1 Closed-Loop Stability: Reducing Subspaces

Let $\varepsilon>0$ and consider the $\alpha$-dimensional compensator (3.1) in closed-loop with the infinite-dimensional DPS (5.1) where $H_N$ and $H_R$ are reducing subspaces. Let the closed-loop state $\omega$ in $H_c \equiv H_1 \times H_2$ be given by

$$\omega = \begin{bmatrix} \omega_1 \\ \omega_2 \end{bmatrix}.$$

where $\omega_1 \equiv \begin{bmatrix} v_N \\ z \end{bmatrix}$ in $H_1 \equiv H_N \times R^\alpha$ and $\omega_2 \equiv v_R$ in $H_2 \equiv H_R$. The norm on $H_c$ is defined by

$$||\omega|| = (||\omega_1||^2 + ||\omega_2||^2)^{1/2} \tag{5.3}$$

where $||\omega_1|| = (||v_N||^2 + ||z||^2)^{1/2}$ (5.4)

From (3.1) and (5.1), the closed-loop state satisfies (4.1) with

$$A_{11} = \begin{bmatrix} A_N+B_NL_{11}C_N & B_NL_{12} \\ L_{21}\ C_N & L_{22} \end{bmatrix} \tag{5.5a}$$

$$A_{12} = \begin{bmatrix} B_NL_{11}C_R \\ L_{21}C_R \end{bmatrix} \tag{5.5b}$$

$$A_{21} = \begin{bmatrix} B_RL_{11}C_N & B_RL_{12} \end{bmatrix} \tag{5.5c}$$

$$A_{22} = A_R+B_RL_{11}C_R \tag{5.5d}$$

We want to apply Theo. 2 to (4.1) and (5.5). Clearly $A_{11}$, $A_{12}$, and $A_{21}$ are bounded linear operators. Since $A_R$ generates an exponentially stable $C_0$-semigroup and $B_RL_{11}C_R$ is a bounded operator, $A_{22}$ also generates an exponentially stable $C_0$-semigroup $U_2(t)$ satisfying (4.3) with

$$\begin{cases} K_2 = K_R \\ \sigma_2 = \sigma_R-K_R||B_RL_{11}C_R|| \end{cases} \tag{5.6}$$

<u>when</u> $||L_{11}||$ <u>is sufficiently small</u>, i.e.,

$$||L_{11}|| < \frac{\sigma_R}{K_R\ ||B_R||\ ||C_R||} \tag{5.7}$$

This can be satisfied when the spillover terms $B_R$ and $C_R$ are not too large. Finally, we must show that the bounded operator

$$A_1=A_{11}-A_{12}A_{22}^{-1}A_{21}$$

generates an exponentially stable $C_o$-semigroup, i.e., has all of its spectrum in the open left-half of the complex plane. However, it turns out that $A_1$ is completely determined by the synthesis of Sec. 3.0: this is shown in the next lemma:

Lemma 3: $A_1 = F_N = \bar{A}_N+\bar{B}_N\bar{L}\ \bar{C}_N$ (5.8)

when reducing subspaces are used in Sec. 3.0 and (5.7) is satisfied. The proof of this lemma uses Lemma 2, and it is given in [24]. Since $A_1=F_N$ and this is designed in Sec. 3.0 with any desired stability (i.e., eigenvalues of $F_N$ are arbitrarily close to any desired locations in the complex plane), we have $U_1(t) = e^{F_N t}$ and

$$||e^{F_N t}|| \leq K_N e^{-\sigma_N t}, \ t \geq 0 \tag{5.9}$$

where $K_N \geq 1$ and $\sigma_N > 0$.

Now Theo. 2 can be applied and we summarize the above discussion as:

Theorem 4: Assume

(a) a singular perturbations model reduction (Sec. 2.0) of the DPS (1.1) exists with reducing subspaces (2.4);

(b) the stable controller-compensator synthesis condition (3.5) of Theo. 1 is satisfied for the reduced system $(A_N,B_N,C_N,\tilde{D}_N)$ when $\varepsilon=0$;

(c) the control and observation spillover are sufficiently small that (5.7) holds.

Then there exists an $\varepsilon_o$, bounded above by (4.12) with $(\sigma_1,K_1)$ given by

$(\sigma_N, K_N)$ in (5.9) and $(\sigma_2, K_2)$ given in (5.6), such that any $\alpha$-dimensional controller-compensator (3.1) synthesized for stable control when $\varepsilon=0$ will stabilize the DPS (1.1) i.e., the closed-loop system will be exponentially stable for any $0<\varepsilon\leq\varepsilon_o$.

5.2 Closed-Loop Stability: Nonreducing Subspaces

Now the general case of closed-loop stability where $H_N$ and $H_R$ are nonreducing subspaces can be considered. Let $\omega$ be the closed-loop state of the controller-compensator (3.1) with the general DPS (2.3) and keep the same definitions of $\omega_1, \omega_2$, $H_1, H_2$, and the norms (5.3)-(5.4). In this case, for $\varepsilon>0$, the closed-loop system satisfies:

$$\begin{cases} \dfrac{\partial \omega_1}{\partial t} = \tilde{A}_{11}\omega_1 + \tilde{A}_{12}\omega_2 \\ \varepsilon\dfrac{\partial \omega_2}{\partial t} = \tilde{A}_{21}\omega_1 + \tilde{A}_{22}\omega_2 \end{cases} \tag{5.10}$$

where, by using (5.5), we have

$$\tilde{A}_{11} = A_{11},\ \tilde{A}_{22} = A_{22},\ \text{and}\ \tilde{A}_{12} = A_{12} + \Delta A_{12},\ \text{and}\ \tilde{A}_{21} = A_{21} + \Delta A_{12}$$

with

$$\begin{cases} \Delta A_{12} \equiv \begin{bmatrix} A_{NR} \\ 0 \end{bmatrix} \\ \Delta A_{21} \equiv \begin{bmatrix} A_{RN} & 0 \end{bmatrix} \end{cases} \tag{5.11}$$

Again, we want to apply Theo. 2. Since $\tilde{A}_{11}=A_{11}$ and $\tilde{A}_{22}=A_{22}$ and $\tilde{A}_{11}$, $\tilde{A}_{12}$, $\tilde{A}_{21}$ are bounded operators (because $A_{RN}$ and $A_{NR}$ are finite-rank operators), we need only check the stability of

$$\tilde{A}_1 = \tilde{A}_{11}-\tilde{A}_{12}\tilde{A}_{22}^{-1}\tilde{A}_{21}$$

$$= A_{11} - (A_{12}+\Delta A_{12})\ A_{22}^{-1}(A_{21}+\Delta A_{21})$$

$$= A_1 - (A_{12}A_{22}^{-1}\Delta A_{21}+\Delta A_{12}A_{22}^{-1}A_{21} + \Delta A_{12}A_{22}^{-1}\Delta A_{21})$$

$$= A_1 - \Delta A_1 \tag{5.12}$$

with

$$\Delta A_1 \equiv \begin{bmatrix} Q_{11} & Q_{12} \\ Q_{21} & 0 \end{bmatrix} \tag{5.13}$$

with

$$\left\{ \begin{array}{llll} Q_{11} = Q_A+Q_B+Q_C & & Q_A & = B_N L_{11} C_R A_{22}^{-1}\ A_{RN} \\ Q_{12} = A_{NR}A_{22}^{-1}\ B_R\ L_{12} & & Q_B & = A_{NR}A_{22}^{-1}B_R L_{11}\ C_N \\ Q_{21} = L_{21}C_R A_{22}^{-1}\ A_{RN} & & Q_C & = A_{NR}A_{22}^{-1}\ A_{RN} \end{array} \right. \tag{5.14}$$

The following lemma completes the analysis of the general case:

<u>Lemma 4</u>: If (5.7) is satisfied, then

$$\tilde{F}_N = \begin{bmatrix} \tilde{A}_N+\tilde{B}_N\bar{L}_{11}\tilde{C}_N & \tilde{B}_N\bar{L}_{12} \\ \bar{L}_{21}\tilde{C}_N & \bar{L}_{22} \end{bmatrix} \quad \text{in (3.2)}$$

can be written

$$\tilde{F}_N = F_N-\Delta F_N \tag{5.15}$$

where $F_N$ is the same as in Lemma 3 and

$$\Delta F_N = \begin{bmatrix} H_{11} & H_{12} \\ H_{21} & 0 \end{bmatrix} \tag{5.16}$$

with

$$\begin{cases} H_{11} = H_A + H_B + H_C & H_A = B_N \bar{L}_{11} C_R A_R^{-1} A_{RN} \\ H_{12} = A_{NR} A_R^{-1} B_R \bar{L}_{12} & H_B = A_{NR} A_R^{-1} B_R \bar{L}_{11} C_N \\ H_{21} = \bar{L}_{21} C_R A_R^{-1} A_{RN} & H_C = A_{NR}(A_R^{-1} - A_R^{-1} B_R \bar{L}_{11} C_R A_R^{-1}) A_{RN} \end{cases} \tag{5.17}$$

Furthermore $\Delta F_N = \Delta A_1$ or, equivalently, $\tilde{F}_N = \tilde{A}_1$ (5.18)

The proof of this lemma is given in [24]. Consequently, $\tilde{F}_N = \tilde{A}_1$ can be synthesized as in Sec. 3.0 so that it generates the exponentially stable $C_o$-semigroup $\tilde{U}_1(t) = e^{\tilde{F}_N t}$ satisfying:

$$||e^{\tilde{F}_N t}|| \leq K_N e^{-\sigma_N t}, \quad t \geq 0 \tag{5.19}$$

where $K_N \geq 1$ and $\sigma_N > 0$. Note these numbers may be different from the ones in (5.9); in particular, $K_N$ may differ even though $\sigma_N$ is kept the same.

Now Theo. 2 may be applied and our general closed-loop stability result summarized as:

Theorem 5: Assume

(a) a singular perturbations model reduction (Sec. 2.0) of the DPS (1.1) exists for some pair of subspaces $H_N$ and $H_R$;

(b) the stable controller-compensator condition (3.5) of Theo. 1 is satisfied for the reduced system $(\tilde{A}_N, \tilde{B}_N, \tilde{C}_N, \tilde{D}_N)$ in (2.5) when $\varepsilon = 0$;

(c) the control and observation spillover are sufficiently small that (5.7) holds.

Then there exists an $\varepsilon_o$, bounded above by (4.12) with $(\sigma_1, K_1)$ given by $(\sigma_N, K_N)$ in (5.19) and $(\sigma_2, K_2)$ given in (5.6), such that any $\alpha$-dimensional controller-compensator (3.1) synthesized for stable control when $\varepsilon = 0$ will stabilize the DPS (1.1), i.e. the closed-loop system will be exponentially stable, for any $0 < \varepsilon \leq \varepsilon_o$.

The size of $\varepsilon_o$ is directly related to the bounds $M_1$ through $M_6$. These bounds are functions of the modeling error, the control or observation spillover and the residual subsystem stability of a particular model reduction for (1.1).

## 6.0 CONCLUSIONS AND RECOMMENDATIONS

Our main result (Theo. 5 ) gives conditions under which a general finite-dimensional controller-compensator, based on a singular perturbation reduced-order model, will stabilize an infinite dimensional (linear) distributed parameter system. This is extremely useful since all practical feedback compensators must be finite-dimensional in order to be implementable with on-line digital computers. The results are valid for large-scale lumped-parameter systems, as well. They depend on an infinite-dimensional version of the Klimushchev-Krasovskii lemma which we have stated here as Theo. 2.

The most difficult assumption to satisfy is the choice of reduced-order subspace $H_N$ to achieve a singularly perturbed model reduction (3.1)-(3.2) of the distributed parameter system (1.1); the other assumptions are reasonably natural or easy to satisfy. Modal methods have worked for mechanically flexible structures, e.g. [11]. A general discussion of this modeling difficulty is given in [23]; it is a fundamental problem in all large-scale or distributed parameter systems control problems and should not be overlooked. In fact, there may be more than one acceptable choice of $H_N$.

Most singular perturbation methods yield results for "small enough $\varepsilon>0$" based on system behavior when $\varepsilon=0$. Ours are no exception; however, in the style of [21]-[22] for finite-dimensions, we have established an upper bound $\varepsilon_0$ on the size of $\varepsilon$ for infinite-dimensional systems. This bound gives some idea of how small is "small enough" for the stability results to hold. Since there may be several ways to achieve the singularly perturbed model reduction, e.g., time, frequency, or mass scaling for flexible structures [11], each one should be assessed to see how large the upper bound $\varepsilon_o$ can be. The size of $\varepsilon_o$ is related to the amount of modeling error and/or control and observation spillover present in the model reduction.

Furthermore, it is extremely important to be able to relate the small parameter $\varepsilon$ to physical quantities in the system, in order to be able to make use of the stability results. For example, $\varepsilon$ can be related to the spectral separation of high and low frequencies in some flexible structures [11]; when the separation is adequate to guarantee that $0<\varepsilon\leq\varepsilon_o$, there is reasonable theoretical assurance of stability for infinite-dimensional distributed parameter systems.

Singular perturbations approaches yield different stability bounds for distributed parameter systems than those obtained through regular perturbations (e.g. [7]). However, a better knowledge of the residual subsystem parameters is generally required for the calculation of the reduced system parameters (2.5) when singular perturbation methods are to be used unsuccessfully. A comparison of stability bounds obtained via singular and regular perturbations is warranted for the variety of interesting distributed parameter problems under consideration in the literature.

ACKNOWLEDGEMENT

This work was partially supported by the National Science Foundation under grant number ECS-80-16173 and the National Aeronautics and Space Administration under grant number NAG-1-171 and contract number NAS-9-16053. Any opinions, findings, and conclusions or recommendations expressed in this publication are those of the author and do not necessarily reflect the views of NSF or NASA.

REFERENCES

1. Lions, J.L., Some Aspects of the Optimal Control of DPS, SIAM, Philadelphia, PA, 1972.

2. Ray, W.H., and Lainiotis, D., eds., DPS: Identification, Estimation, and Control, M. Dekker, NY , 1978.

3. Russell, D., "Controllability & Stabilizability Theory for Linear PDE: Recent Progress & Open Questions", SIAM Review, 20, 371-388, 1978.

4. Curtain, R., and Pritchard, A., Infinite Dimensional Systems Theory, Springer, NY, 1978.

5. Aziz, A., et.al.eds., Control Theory of Systems Governed by PDE, Academic Press, NY, 1977.

6. Balas, M., "Trends in Large Space Structure Control Theory: Fondest Hopes, Wildest Dreams", IEEE Trans. Autom. Control, Vol. AC-27, 522-535, 1982.

7. Balas, M., "Toward A (More) Practical Theory of Distributed Parameter System Control", Advances in Dynamics & Control: Theory and Applications, Vol. 18, C.T. Leondes, editor, Academic Press, NY, 1982.

8. Kokotovic, P., O'Malley, R., and Sannuti, P., "Singular Perturbations & Order Reduction in Control Theory: An Overview", Automatica, 12, 123-132, 1976.

9. Balas, M.,"Singular Perturbations & Model Reduction Methods for Large-Scale and Distributed Parameter Systems,"Proc. of 17th Annual Allerton Conference on Communication, Control, and Computing, Monticello, IL, 434-448, 1979.

10. Balas, M.,"Reduced-Order Feedback Control of DPS Via Singular Perturbation Methods,"J. Math. Analysis & Appl., Vol. 87, 281-294, 1982.

11. Balas, M., "Closed-Loop Stability of Large Space Structures Via Singular & Regular Perturbation Techniques: New Results", Proc. of IEEE Control & Decision Conf., Albuquerque, NM, 1980.

12. Kato, T., Perturbation Theory For Linear Operators, Springer, NY, 1966.

13. Kwakernaak, H. & Sivan, R., Linear Optimal Control Systems, J.Wiley & Sons, NY, 1972.

14. Curtain, R. & Pritchard, A., Functional Analysis & Modern Applied Mathematics, Academic Press, NY, 1977.

15. Walker, J., Dynamical Systems & Evolution Equations: Theory & Applications, Plenum Press, NY, 1980.

16. Kimura, H., "Pole Assignment by Gain Output Feedback", IEEE Trans. Autom. Control, AC-20, 509-516, 1975.

17. Davison, E., and Wang, S., "On Pole Assignment in Linear Multivariable Systems Using Output Feedback", Ibid, 516-518, 1975.

18. Vidyasagar, M., Nonlinear Systems Analysis, Prentice-Hall, NJ, 1978.

19. Klimushchev, A., and Krasovskii, N., "Uniform Asymptotic Stability of Systems of Differential Equations With A Small Parameter in the Derivative Terms", J. Appl. Math & Mech., Vol. 25, 1961.

20. Wilde, R., and Kokotovic, P., "Stability of Singularly Perturbed Systems & Networks With Parasitics", IEEE Trans. Autom. Contr., Vol. AC-17, 616-625, 1973.

21. Zien, L., "An Upper Bound For The Singular Parameter in a Stable Singularly Perturbed System", J. Franklin Inst. Vol. 295, 373-381, 1973.

22. Javid, S.H., "Uniform Asymptotic Stability of Linear Time-Varying Singularly Perturbed Systems", Ibid, Vol. 305, 27-37, 1978.

23. Kokotovic, P., "Subsystems, Time-Scales, & Multi-Modeling", Proc. of IFAC Symp. on Large Scale Systems, Toulouse, France, 1980.

24. Balas, M., "Stability of Distributed Parameter Systems With Finite-Dimensional Controller-Compensators Using Singular Perturbations", J. Math Analysis & Appl. (to appear).

# TRANSITION LAYERS, ANGULAR LIMITING SOLUTIONS AND INTERNAL LAYERS IN SINGULARLY PERTURBED NONLINEAR EIGENVALUE PROBLEMS

Claude M. Brauner†
Departement de Mathematiques
- Informatique - Systemes
Ecole Centrale de Lyon
69130 Ecully, France
and
Basil Nicolaenko
Center for Nonlinear Studies, MS B258
Los Alamos National Laboratory
Los Alamos, NM 87545 USA

In this review article we shall survey recent developments in singular perturbations of nonlinear eigenvalue problems (N.L.E.P.). Nonlinear eigenvalue problems in bounded domains characteristically exhibit multiple solutions; as the eigenvalue parameter varies, connected components of solutions go through bending points (turning points), which mathematically are closely related to bifurcations. Singularly perturbed nonlinear eigenvalue problems exhibit a wealth of singular limits; including boundary layers, angular limiting solutions, transition layers, internal layers, and free boundary limits. In the first section, after reviewing basic facts for N.L.E.P.'s, we will survey the asymptotic behavior of branches of

† Consultant, Los Alamos National Laboratory.

solutions as a nonlinear eigenvalue parameter goes to infinity. In the second section, we will emphasize the close feedback and coupling between the generic multiplicity of solutions of N.L.E.P.'s and the corresponding multiplicity of singular limits. Typically, the singular limit problems are free boundary problems (F.B.P.) with internal layers centered on the a priori unknown free boundaries; moreover, the free boundaries themselves can exhibit multiplicity, with their bending points (bifurcations) not necessarily related to bending points of the original perturbed N.L.E.P. The plan of the article is as follows:

<u>Section 1</u>: Asymptotic study of N.L.E.P.

1.1 Review of basic facts for N.L.E.P.

1.2 Asymptotic behavior of a N.L.E.P. as a singular perturbation problem: the local point of view

1.3 Global branch tracking results.

<u>Section 2</u>. N.L.E.P. whose singular limits exhibit internal layers and angular points

2.0 Introduction

2.1 Transition Layers and Angular Limiting Solutions

2.2 Angular Limiting Solutions as end points of branches of solutions

2.3 A special type of Angular Limiting Solution: A Free Boundary Problem

## SECTION 1. ASYMPTOTIC STUDY OF NONLINEAR EIGENVALUE PROBLEMS

### 1.1 Review of Basic Facts for N.L.E.P.

A nonlinear eigenvalue problem is a nonlinear elliptic problem parameterized by an eigenvalue $\lambda$ [1-4]:

$$\begin{cases} Au = \lambda g(x,u) \text{ in } \Omega \in R^n \quad , n \geq 1 \quad , \\ \lambda \in R^+ \quad ; \text{ plus boundary conditions on } \partial\Omega \quad ; \\ Au = \begin{cases} -\Delta u \\ \text{or} \quad -\sum_{i,j=1}^{3} \frac{\partial}{\partial x_i} a_{ij}(x) \frac{\partial u}{\partial x_j} \quad ; \end{cases} \end{cases} \tag{1.1}$$

the usual boundary conditions are: $u/\partial\Omega = 0$ or $u$ given $\geq 0$. We now specify the hypothesis on $g(x,u)$: $\Omega \times R \to R$ (we note that $g(x,u)$ can also depend upon a perturbation parameter $\varepsilon$):

Hypothesis (H1):

(i) $g(x,u)$ is continuous in $x$ and $u$, $C^1$ in $u$, smoother if necessary;

(ii) $g(x,0) > 0$, $\forall\, x \in \Omega$.

Such hypothesis excludes the trivial solution $u \equiv 0 \;\forall\, \lambda$.

Hypothesis (H2):

$g(x,u) \geq 0, \forall u \geq 0.$

The results described here can be extended to more general $g(x,u,\nabla u)$.

Proposition 1.1. [4]

Under hypothesis (H1), the connected component C to which ($\lambda = 0$, $u = 0$) belongs:

(i) consists of positive solutions

(ii) is unbounded in $R^+ \times C^1(\bar{\Omega})$.

Typical cases are described in Figures 1 and 2.

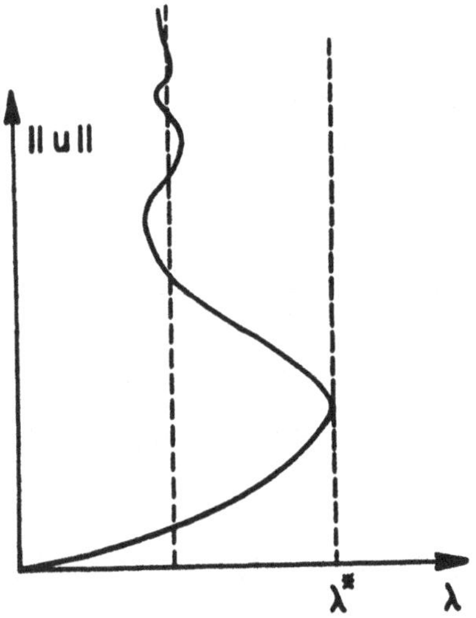


Figure 1

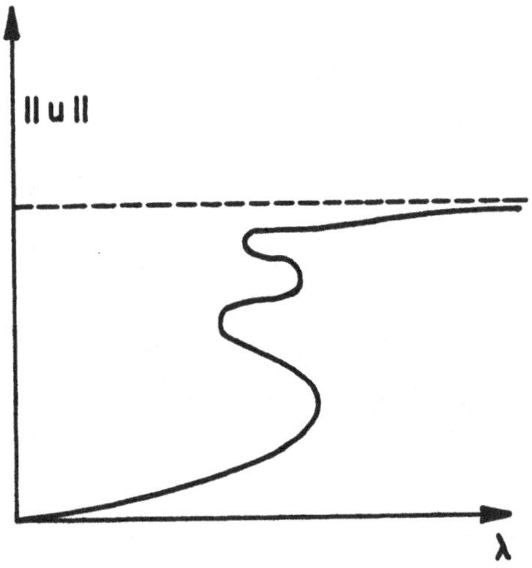


Figure 2

In Fig. 1 the connected component C is bounded in $\lambda$, but unbounded in the functional space of u; there exist no classical solutions beyond some critical $\lambda^*$. In Fig. 2 the connected component C is bounded in the u norm and therefore unbounded in the parameter $\lambda$; C goes to $\infty$ in $\lambda$. Now denoting by $g_u(\lambda,u)$ the partial derivative of g w.r.t. u:

Proposition 1.2.

Under hypothesis (H1) and (H2) the minimal branch $\underline{u}(\lambda)$ starting at $(\lambda = 0,\ u = 0)$ has a maximal continuation such that $A - \lambda g_u(\lambda,u)I$ is invertible; $\underline{u}(\lambda)$ is monotone increasing in $\lambda$ on that branch.

We now introduce the notion of Bending Point:

Definition. $(\lambda_B,u_B)$ is a bending point if there exists a small neighborhood B of $\lambda_B$ in $R^+$, such that for $\forall\ \lambda < \lambda_B$ in B (or $\forall\ \lambda > \lambda_B$), $\exists$ two distinct solutions $u_1(\lambda)$ and $u_2(\lambda)$; and for $\forall\ \lambda > \lambda_B$ in B (or respectively $\forall\ \lambda < \lambda_B$), $\exists$ no solution of the N.L.E.P.; a necessary condition for a bending point is:

$$\ker\{A - \lambda_b g_u(\lambda_B,u_B)I\} \neq \emptyset \quad .$$

This condition means that the linearization of problem (1.1) at $(\lambda_B,u_B)$ is not invertible. This is a necessary, but by no means sufficient condition for bifurcation at $\lambda_B$. For discussion of sufficient conditions we refer to [1-2]; in particular a simple case corresponds to a one dimensional kernel for the linearization.

The case of Fig. 1 is generic if the following hypothesis is verified:

Hypothesis (H3):

$g = g(u)$ satisfies (H1) + (H2) and $g$ is monotone increasing, convex.

Proposition 1.3. [3-4]

Under Hypothesis (H3)

(i) there exists a critical $\lambda^*$, such that for $\lambda > \lambda^*$ there exists no classical solution of the N.L.E.P..

(ii) as $\lambda \uparrow \lambda^*$, $\underline{u}(\lambda) \uparrow u^*(\lambda^*)$ in $H_0^1(\Omega)$ weak, where $u^*(\lambda^*)$ is a weak solution of the N.L.E.P. (1.1).

An example which verifies Proposition 1.3 is the famous Gelfand problem [3]:

$$-\Delta u = \lambda e^u, \; u = 0/\partial\Omega. \tag{1.2}$$

In Fig. 3 we have plotted the component C of Proposition 1.1 when the domains are balls in $R^n$.

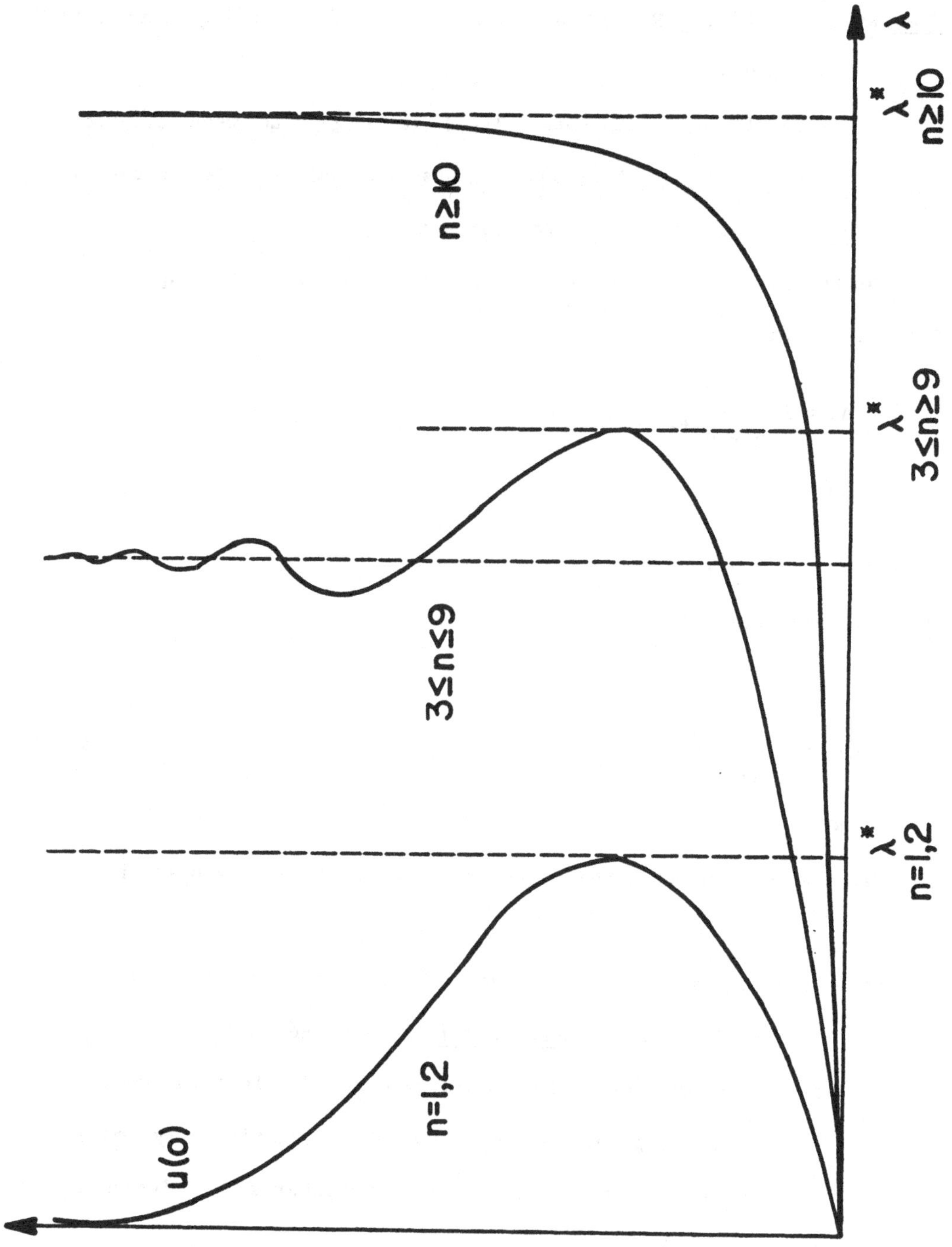


Figure 3

<u>Remarks</u>: (i) If $n \leq 9$ then $\lambda^*$ is also a regular bending point for $\underline{u}(\lambda)$.

(ii) If $n \geq 10$ then $u^*(\lambda^*)$ is a genuine weak solution (blow-up of u(0)). We observe that we have a singular endpoint of the regular arc.

We now give an example with a nonconvex g(u) which corresponds to the case of Fig. 2 [5-7]:

$$\begin{cases} \Delta u = \lambda \dfrac{u^m}{\eta + u^{m+k}} \\ u/\partial\Omega = 1 \\ \eta > 0 \quad , m \geq 1 \quad , 0 < k < 1 \quad , \end{cases} \tag{1.3}$$

which is equivalent to:

$$\begin{cases} u = 1-v \\ -\Delta v = \lambda \dfrac{(1-v)^m}{\eta+(1-v)^{m+k}} \\ v/\partial\Omega = 0 \quad . \end{cases} \tag{1.4}$$

In Fig. 4 we have plotted the component C in the case of a disk in $R^2$.

<u>Remarks</u>: (i) The component C of positive solutions $0 \leq v < 1$ (i.e., $0 < u \leq 1$) is <u>unbounded in λ</u>, bounded in u.

(ii) The component C has an arbitrarily large number of bending points as one takes $\eta > 0$ sufficiently small! (This is true only for hyperspheres of dimension $n \geq 2$.)

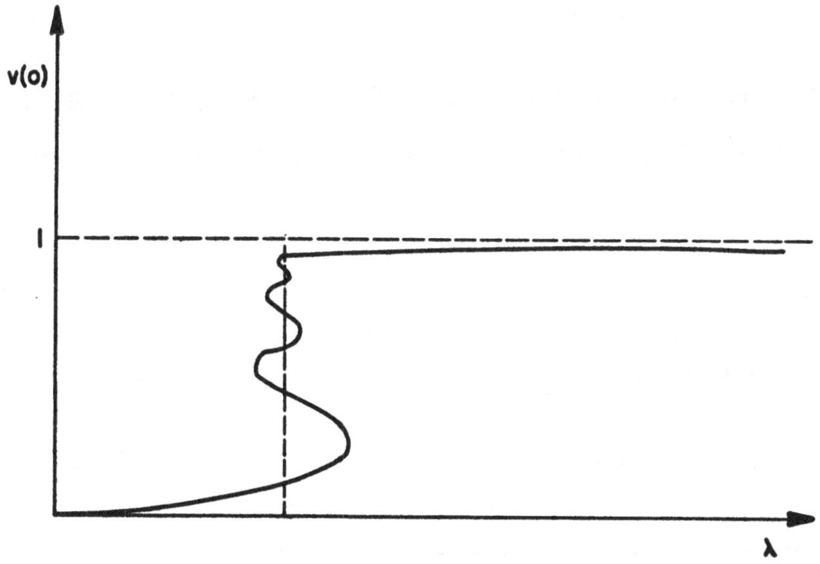


Figure 4

This leads us to the asymptotic problem considered in the next section. As $\lambda \to +\infty$, we can set $\lambda = 1/\varepsilon$, $\varepsilon \to 0^+$. The problem is then to track the branches $v(\lambda)$. We claim that $v(\lambda) \to 1$ in $L^p(\Omega)$ strong with a boundary layer (loss of the boundary condition $v/\partial\Omega = 0$).

## 1.2 Asymptotic Behavior of a N.L.E.P. as a Singular Perturbation Problem: The Local Point of View.

Let $g(x,u)$ as in Sec. 1.1, satisfying (H1) or (H2). We consider the N.L.E.P.

$$Au = \lambda g(x,u) \quad , \; u/\partial\Omega = 0 \tag{1.5}$$

and we are interested in the large values of $\lambda$, provided that the projection on $R^+$ of the connected component C is unbounded. Of course it is convenient to set

$$\varepsilon = 1/\lambda \quad , \; \varepsilon > 0 \text{ small} \quad , \tag{1.6}$$

and to rewrite (1.5) as

$$\varepsilon Au = g(x,u) \quad , \; u/\partial\Omega = 0 \qquad \text{(1.5)bis}$$

but the notation "$\varepsilon$" has the disadvantage of hiding the "bifurcation" aspect of the problem.

The classical - or "local" - point of view for studying solutions $u_\varepsilon(x)$ of (1.5)bis as $\varepsilon \to 0$ is the following: the reduced problem (i.e., (1.5)bis when $\varepsilon$ is set equal to zero):

$$g(x,u) = 0 \tag{1.7}$$

is assumed to have a smooth solution $u_0(x)$, and the solutions of (1.5)bis are studied relative to $u_0$. The solution $u_0$ is supposed to be stable in the following sense:

$$\frac{\partial g}{\partial u}(x,u_0(x)) \geq \gamma > 0 \quad \forall\, x \in \bar{\Omega} \quad . \tag{1.8}$$

As far as layer terms are concerned, a hypothesis of stability in the boundary layer must be introduced; for $x \in \partial\Omega$ fixed, set

$$\bar{g}(x,t) = h(x,u_0(x) + t) - h(x,u_0(x)) \quad . \tag{1.9}$$

It is assumed:

$$\int_0^\xi \bar{g}(x,t)\,dt > 0 \quad \begin{array}{ll} \forall\, \xi \in (0, -u_0(x)] & \text{if } u_0(x) < 0 \quad , \\ \forall\, \xi \in [-u_0(x), 0) & \text{if } u_0(x) > 0 \quad . \end{array} \tag{1.10}$$

The set of conditions (1.8) (1.10) is known as Fife's conditions.

Basic references for this "local" point of view are found in many previous publications BRISH [8], BOGLAEV [9], FIFE [10], and DE VILLIERS [11] (who extended simultaneously and independently BERGER'S and FRAENKEL's results [12]), VAN HARTEN [13], several papers of FRED HOWES (see the survey [14]), ECKHAUS's book [15], ....

In dimension $n = 1$, the concept of stability of $u_0$ can be better understood by writing (1.5)bis as a first order nonlinear system (VASILEVA [16], O'MALLEY [17]).

Under the above hypothesis, a typical result is the following: there exists a family of solutions $u_\varepsilon(x)$ of (1.5)*bis which* converges to $u_0$ uniformly on closed subsets of $\Omega$. Either the functions $u(x,\varepsilon)$ are constructed via a uniformly valid asymptotic expansion, or an argument of upper and lower solutions is used. It is clear that this result does not give any information on the deep structure of the branches of solutions of the N.L.E.P. (1.5) for $\lambda$ large. The next section will partially answer the problem.

### 1.3 Global Branch Tracking Results

In a series of papers, [18-20] Clement and Peletier have obtained the global behavior of the connected component C defined in Proposition 1.1 as $\lambda \to +\infty$. Essentially, they show that under Hypothesis (H1) + (H2), all bounded superharmonic solutions of a N.L.E.P. converge to a "minimal" solution of the reduced equation

$$g(x,u) = 0 \quad ,$$

as $\lambda \to +\infty$.

Proposition 1.4. [20]

With the hypothesis (H1) + (H2) and if:

$(\lambda,u) \in C \Rightarrow \sup v(x) < M \quad , x \in \Omega \quad ,$

Then:

1) $\bar{u}(x) \overset{\text{def.}}{=} \sup(u(x):(\lambda,u) \in C)$, for fixed $x \in \Omega$,

is positive, bounded and <u>superharmonic</u>.

2) $g(x,\bar{u}(x)) = 0$ a.e. in $\Omega$.

3) $\bar{u}(x) = \inf \{v:\Omega \to R | v \geq 0$, v superharmonic and $g(x,v(x)) = 0$ a.e. in $\Omega\}$.

4) $\int |\underline{u}(\lambda) - \bar{u}|^2 dx \to 0$ as $\lambda \to \infty$,

where $\underline{u}(\lambda)$ is the minimal, positive superharmonic function such that $(\lambda, \underline{u}(\lambda)) \in C$.

5) Let $u_o$ be the principal eigenfunction of $-\Delta$:

$$\begin{cases} -\Delta u_o = \lambda_o u_o \\ u_o = 0/\partial\Omega \\ u_o(x) \geq 0 \quad , \sup_{\Omega} u_o(x) = 1 \quad . \end{cases}$$

Then

6) $\int_{\Omega} u_o |\nabla\bar{u}|^2 \, dx < \infty$

7) $\lim_{\lambda\to\infty} \int_{\Omega} (u_o)^2 |\nabla\bar{u} - \nabla\underline{u}(\lambda)| \, dx = 0$

The above convergence estimates reflect the presence of a boundary layer on $\partial\Omega$; they involve weighted Sobolev norms with the weight function $u_o$ null on $\partial\Omega$.

Remark:

These results are extended by Clement and Peletier [18-20] to the more general case:

$Au = \lambda g(x,u,\nabla u)$.

## SECTION 2: NONLINEAR EIGENVALUE PROBLEMS WHOSE SINGULAR LIMITS EXHIBIT ANGULAR POINTS AND INTERNAL LAYERS

### 2.0 Introduction

In this section we shall consider a generic situation for a class of perturbed nonlinear eigenvalue problems:

$$P_\varepsilon(\lambda,u) \equiv 0 \quad ,$$

such that:

(1) $P_\varepsilon(\lambda,u) = 0$ depends upon two parameters $\varepsilon$ and $\lambda$ and has a bounded solution $u_\varepsilon(\lambda)$ $\forall\ \lambda$, $\forall\ \varepsilon > 0$, $\varepsilon \neq 0$.

(2) The Formal Reduced Problem $P_0(\lambda,u) = 0$ has no "acceptable" solutions for $\lambda > \lambda_c$.

By this we mean there are either no classical smooth solutions for $\lambda$ greater than the critical parameter $\lambda_c$ or that there are no smooth solutions satisfying some given constraints, such as positivity. The major problem is: what can we say of the limit of $u_\varepsilon(\lambda)$, $\lambda > \lambda_c$ when $\varepsilon \to 0^+$? The basic dilemma is that such a limit cannot satisfy the formal reduced problem $P_0(\lambda,u) = 0$. In the next paragraphs we will show that such singular limits appear with transition layers, angular limiting solutions and internal layers. Often the internal layers are associated with free boundary value problems (F.B.P.)which are in fact the true singular limit of the perturbed problem $P_\varepsilon(\lambda,u) = 0$. These singular limiting free boundary value problems can exhibit multiplicity and bifurcations themselves; they are nonlinear eigenvalue F.B.P. with bending (turning) points.

## Section 2.1. Transition Layers (T.L.) and Angular Limiting Solutions (A.L.S.)

Within the context of one-dimensional singular perturbation problems, let us recall some definitions.

Consider the problem on the interval $\Omega = (a,b)$

$$\varepsilon u'' = g(x,u) \quad , \ u(a) = A \quad , \ u(b) = B \quad , \tag{2.1}$$

where g is as in Section 1.1.

A Transition Layer (T.L.) is a nonuniformity characterized by an abrupt transition at some interior point $x_0 \in (a,b)$ between distinct solutions of the reduced equation

$$g(x,u) = 0 \quad . \tag{2.2}$$

Transition Layers are called "Shock Layers" by some authors because the nonuniformity might be considered roughly reminiscent of shock phenomena in fluid mechanics.

Transition layers have been studied by FIFE [21,22], FIFE and GREENLEE [23] in higher dimensions, HOWES (see [14] and the list of references therein). The approach of these authors is similar in that the occurence of T.L. is investigated by studying the properties of functionals of the form

$$T(x) = \int_{u_1(x)}^{u_2(x)} h(x,t)\, dt \tag{2.3}$$

where $u_1$, $u_2$ are distinct solutions of the reduced problem (2.2).

As in the case of boundary layers, the concept of stability of solutions of the reduced problem plays a crucial role. We refer to the articles of FRED HOWES, and, again for a link with first order nonlinear systems, to VASILEVA [16, Chapt. II, §2].

Of course, Transition Layers appear in more general nonlinear problems, viz.

$$\varepsilon u'' = g(x,u,u') \quad , \; u(a) = A \quad , \; u(b) = B \quad , \tag{2.4}$$

and it is interesting to remark such T.L. might occur only for very special choices of the data. For example, consider the problem (O'MALLEY [24, p. 9]):

$$\varepsilon u'' = uu' \quad , \; u(-1) = A \quad , \; u(1) = B \quad . \tag{2.5}$$

An elementary computation shows that a T.L. at $x = 0$ appears <u>only</u> if

$A = -B > 0!$ Let $u_\varepsilon$ the (unique) solution of (2.5). Define

$$u_L \equiv A \quad , \; u_R \equiv B \qquad (R = \text{right},\ L = \text{left}) \quad . \tag{2.6}$$

Then

$$u_\varepsilon(x) \to u_o(x) = \begin{cases} u_L(x) \text{ if } -1 \le x < 0 \\ u_R(x) \text{ if } 0 < x \le 1 \quad . \end{cases} \tag{2.7}$$

and remark that there is no Boundary Layer at $x = \pm 1$.

The notion of Angular Limiting Solutions (A.L.S.) has been introduced by HABER and LEVINSON [25]. Roughly speaking, an A.L.S. corresponds to a Transition Layer for the gradients, which appears in the case of essentially nonlinear equations (2.4).

Suppose that the reduced problem associated to (2.4), viz.

$$g(x,u,u') = 0 \tag{2.8}$$

has two solutions $u_L$ and $u_R$, such that $u_L$ and $u_R$ intersect at $x = x_o$ with the conditions

$$u_L(x_o) = u_R(x_o) \quad , \; u_L'(x_o) \neq u_R'(x_o) \quad . \tag{2.9}$$

Then the function $u_o$ defined by

$$u_o(x) = \begin{cases} u_L(x) \quad \text{if} \quad a \le x \le x_o \\ u_R(x) \quad \text{if} \quad x_o \le x \le b \end{cases} \tag{2.10}$$

is called an <u>Angular Limiting Solution</u> of (2.8) and $x_0$ an <u>angular point</u>. If $u_L(a) = A$ and $u_L(b) = B$, **there is** no **Boundary** Layer at $x = a$ and $x = b$.

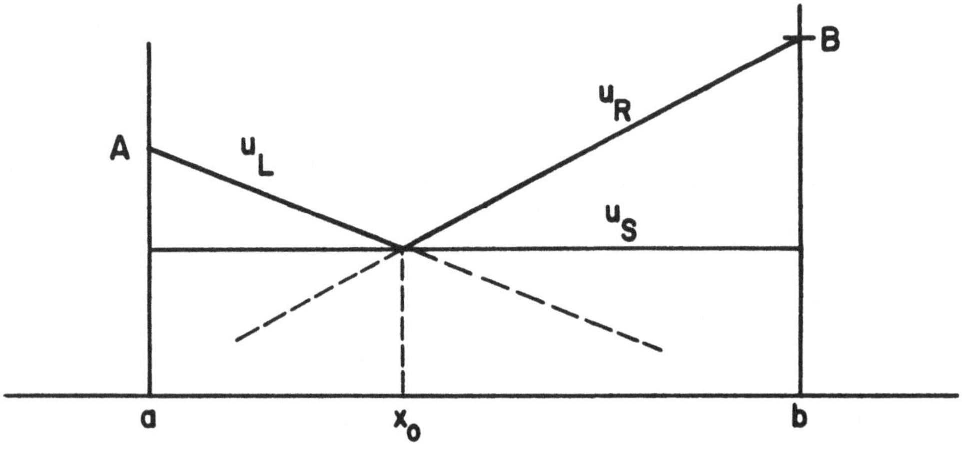


Figure 5

Moreover, $u_L$ and $u_R$ can intersect at $x = x_0$ with a singular solution $u_s$ which does not satisfy the boundary conditions. The stability of $u_L$, $u_R$, $u_s$ has to be discussed. We refer to HOWES [14,26], and, for an asymptotic expansion, to O'MALLEY [27]. VASILEVA [16] considered A.L.S. as discontinuous arcs in the phase-plan.

## 2.2 A.L.S. as Endpoints of Branches of Solutions

In [28-29], the authors considered a substrate reacting inside an enzyme membrane in the case of inhibition by substrate excess. A simplified model is:

$$u'' = \lambda \frac{u}{\varepsilon+u^2} \quad , \; u(-1) = u(1) = 1 \quad . \tag{2.11}$$

For $\varepsilon > 0$, Problem (2.11) admits at least one solution for every $\lambda > 0$, and, for $\lambda_{**}(\varepsilon) < \lambda < \lambda_*(\varepsilon)$, **at least 3** ***solutions***. The reduced problem is

$$u^2u'' = \lambda u \quad , \; u(-1) = u(1) = 1 \tag{2.12}$$

which has 3 solutions for $0 < \lambda < \lambda_*(0)$:

- the 2 solutions of

$$u'' = \frac{\lambda}{u} \quad , \; u(-1) = u(1) = 1 \tag{2.13}$$

denoted by $u_\lambda^1$ and $u_\lambda^2$;

- the singular solution $u_s \equiv 0$ which does not satisfy the boundary conditions.

Now let $\lambda \to 0$. The reduced equation (w.r.t.$\lambda$!) associated to (2.12) is now

$$u^2u'' = 0 \qquad u(-1) = u(1) = 1 \tag{2.14}$$

and we have the convergence results:

Lemma 2.1. As $\lambda \to 0$,

$$\begin{aligned} &u_\lambda^1(x) \to u_0^1(x) = 1 \quad , \\ &u_\lambda^2(x) \to u_0^2(x) = |x| \quad . \end{aligned} \tag{2.15}$$

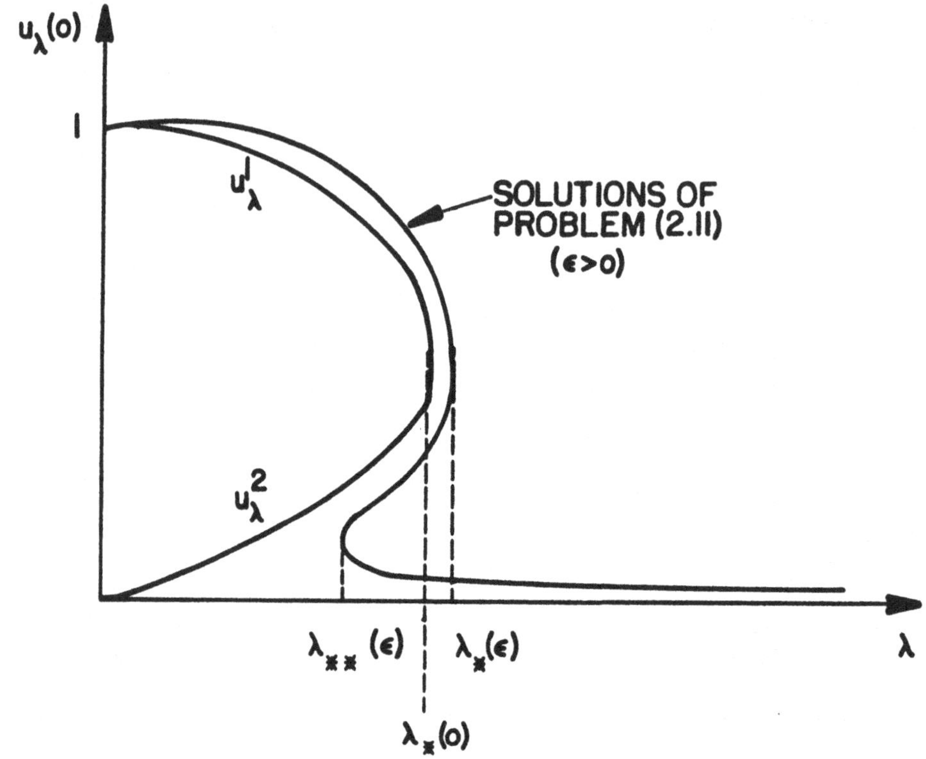


Figure 6

Summarizing, $u_0^1$ is a smooth solution of the reduced problem (2.14), $u_0^2$ is an A.L.S. satisfying the boundary conditions, with x = 0 as an angular point. At x = 0, $u_0^2$ meets the singular solution $u_s \equiv 0$ of (2.14).

From the bifurcation point of view, the A.L.S. ($\lambda = 0$, $u_\lambda = u_0^2$) is an <u>end point</u> of the connected component of positive solutions issued of ($\lambda = 0$, $u_\lambda = 1$). A precise definition of the notion of end point can be found in [7]. This A.L.S. is also the limit of a sequence of regular bending (turning) points $\lambda_{**}(\varepsilon)$ of the perturbed problem (2.11) as $\varepsilon \to 0$.

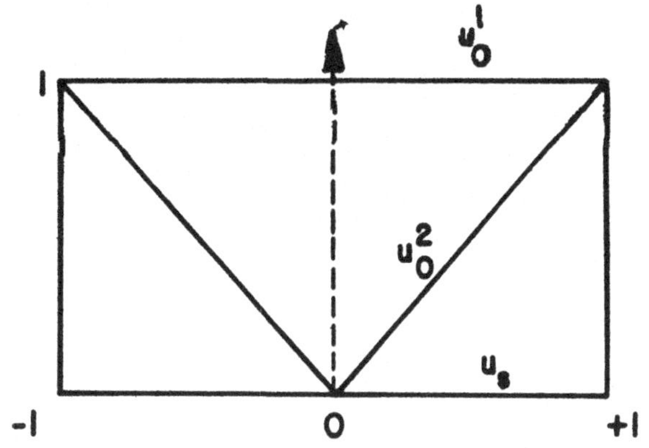


Figure 7

## 2.3 A Special Case of A.L.S.: a Free Boundary Problem

We now consider a simple special case belonging to the general class $P_\varepsilon(\lambda,u)$ which exhibits a free boundary limit [30]:

$$\begin{cases} \Delta u_\varepsilon = \lambda \dfrac{u_\varepsilon}{\varepsilon+|u_\varepsilon|} \quad , \\ u_\varepsilon/\partial\Omega = 1 \quad , \\ \lambda \geq 0 \qquad \varepsilon > 0 \quad ; \end{cases} \tag{2.16}$$

for $\varepsilon$ fixed, $> 0$, the above problem defines a maximal monotone operator; classically it has a unique solution for $\forall\lambda$, $\forall\varepsilon$ fixed:

**Proposition 2.1.**

For $\varepsilon$ fixed $> 0$, $\forall\lambda$, $P_\varepsilon$ has a unique solution $u_\varepsilon(\lambda) \in C_\infty(\bar{\Omega})$, and $0 \leq u_\varepsilon(\lambda) \leq 1$.

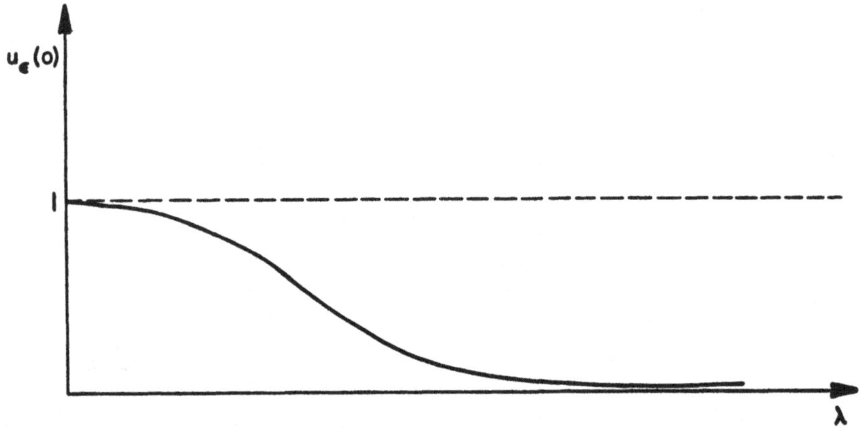


Figure 8

The formal limit problem $P_o(\lambda,u)$ associated to (2.16) is the linear operator:

$$\Delta u = \lambda$$
$$u/\partial\Omega = 1 \tag{2.17}$$

As $\varepsilon \to 0^+$, one can establish that there is a unique, well defined limit $u_o(\lambda)$:

Proposition 2.2. [30]

As $\varepsilon \to 0^+$, $\lambda$ fixed $> 0$, $u_\varepsilon(\lambda) \downarrow u_o(\lambda)$ unique in $C^{1,\alpha}(\bar{\Omega})$ strong, $u_o/\partial\Omega = 1$, $u_o(\lambda) \geq 0$.

We now face the following dilemma: it can easily be established that, as $\lambda$ is greater than some critical value $\lambda_{cr}$, the solutions of the formal limit problem (2.17) admit strictly negative values in some nontrivial open ball. So, $u_o(\lambda)$, which is everywhere $\geq 0$, cannot be a solution of (2.17) for $\lambda > \lambda_{cr}$. In fact, it exhibits a free boundary enclosing some subdomain $\Omega_o \subset \Omega$, where $u_o(\lambda) \equiv 0$. The situation is especially simple when $\Omega$ is a one-dimensional interval.

As $\lambda$ increases, the limit $u_o(\lambda)$ is first $C^\infty$ and satisfies (2.17); but its minimum at $x = 0$ decreases monotonically in $\lambda$. For some $\lambda_{cr}$, the minimum of $u_o(\lambda)$ is exactly zero. As $\lambda > \lambda_{cr}$, an interval $[-a(\lambda), +a(\lambda)]$ appears, where $u_o(\lambda) \equiv 0$. In $[-1, -a(\lambda)]$ and $[+a(\lambda), +1]$, it verifies only locally:

$$+\Delta u_o(\lambda) \equiv \lambda \quad ;$$

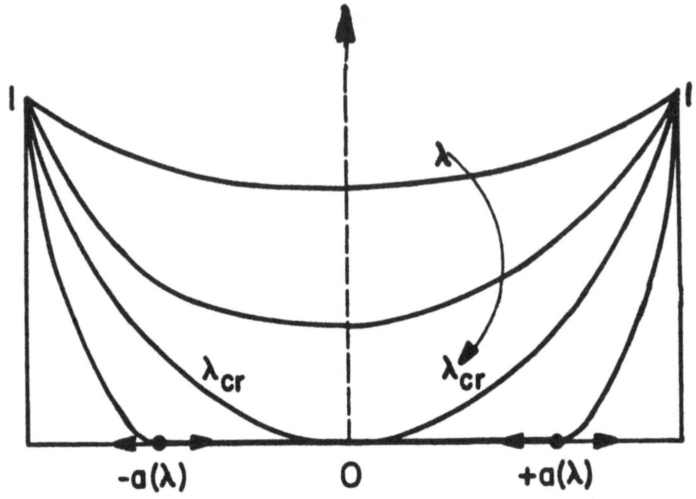


Figure 9

$\pm a(\lambda)$ are angular points, with finite jumps of the Laplacian $\Delta u_o$ (instead of the gradient). Thus, $u_o(\lambda)$ is not $C^2(\bar{\Omega})$. In fact it can be shown that $u_o(\lambda)$ is the unique solution of the Free Boundary Value Problem:

$$\begin{cases} u_o \geq 0 \quad , -\Delta u_o + \lambda \geq 0 \quad , \\ u_o(-\Delta u_o + \lambda) = 0 \text{ in } \Omega \quad , \\ u_o/\partial\Omega = 1 \quad . \end{cases} \tag{2.18}$$

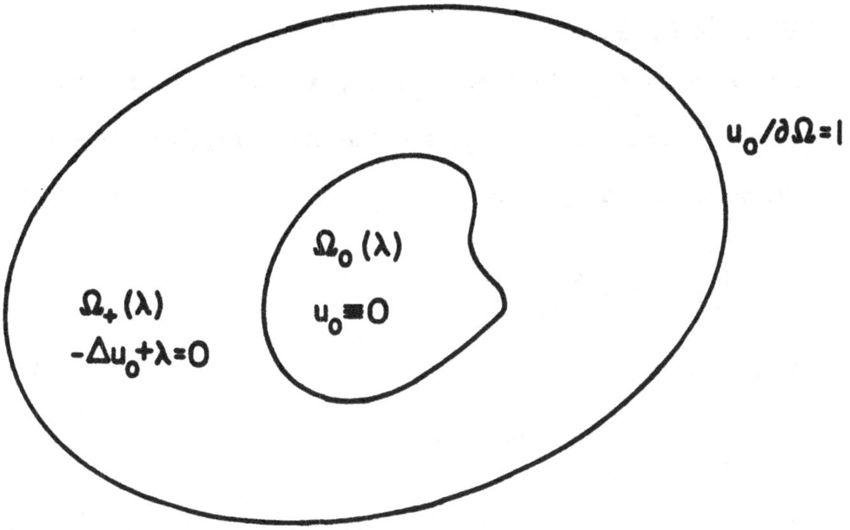


Figure 10

The F.B.P. (2.18) is equivalent to a Variational Inequality (V.I.)

$$(\nabla u_o, \nabla(v-u_o)) + (\lambda, v-u_o) \geq 0 \quad ,$$

$\forall\ v \in K$ , $u_o(\lambda) \in K$ , where K is the convex set defined by

$$K = \left\{ v \in H_o^1(\Omega) \,|\, v \geq 0 \text{ and } v/\partial\Omega = 1 \right\}$$

(i.e., this variational inequality is verified only for those functions v belonging to the constraints set K). For a detailed review of the V.I. formulation of F.B.P., see [31-32].

## 2.4 A N.L.E.P. Whose Singular Limit is a Free Boundary Problem with Multiple Solutions

We now give an example where $P_\varepsilon(\lambda,u)$ is multivalued, the formal limit $P_0(\lambda,u)$ is also multivalued, and the true singular limit problem is a F.B.P. with bending points:

$$P_\varepsilon(\lambda,u) \begin{cases} \Delta u + \lambda \dfrac{A-u}{\varepsilon+|A-u|}\, e^u = 0 \\ u/\partial\Omega = 0 \quad , \quad \varepsilon > 0 \quad , \quad \lambda \geq 0 \quad , \\ A \text{ given constant} > 0 \quad ; \end{cases} \tag{2.19}$$

this problem has been investigated by CONRAD and BRAUNER [33]. First:

Proposition 2.3

For $\forall\varepsilon$ fixed $\geq 0$, $\forall\ \lambda \geq 0$, $P_\varepsilon$ has at least one solution; $\forall$ solution $u_\varepsilon(\lambda)$ verifies

$$0 \leq u_\varepsilon(\lambda) \leq +A \quad ;$$

as $\lambda \to \infty$, $\varepsilon$ fixed, $u_\varepsilon \to +A$ in $L^P(\Omega)$, with stronger convergence results given by Proposition (1.4).

Now the formal limit problem $P_0(\lambda,u)$ is the Gelfand problem, whose properties are summarized in Figure 3:

$$\begin{cases} \Delta u + \lambda e^u = 0 \\ u/\partial\Omega = 0 \quad ; \end{cases} \tag{2.20}$$

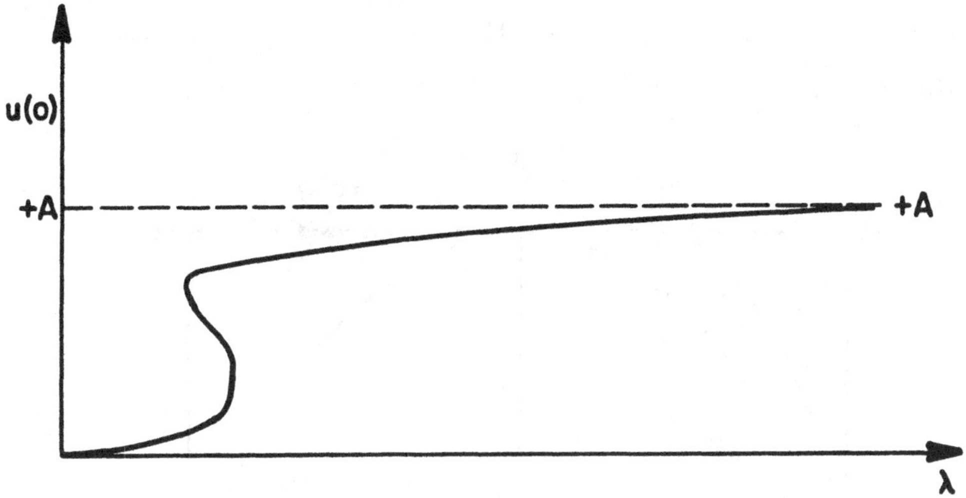


Figure 11

in particular, recall that i) there are no classical solutions to (2.20) for $\lambda > \lambda^*$; ii) the connected component C blows-up in the norm of u. For some $\lambda \le \lambda^*$, there will always exist a solution of $P_o = 0$ with a maximum strictly greater than A. Thus:

Lemma 2.2. Suppose that $\lambda$ is such that either

i) $P_o(\lambda,u) = 0$ has no solutions at all; or

ii) $P_o(\lambda,u) = 0$ has no solutions bounded by +A.

Then the set $\{x \in \Omega | \bar{u}_o(\lambda) = +A\}$ has positive measure, where $\bar{u}_o(\lambda)$ is the limit of $\bar{u}_\varepsilon(\lambda)$, maximal solution of $P_\varepsilon(\lambda,u) = 0$.

Again, the situation is simpler when $\Omega$ is a one-dimensional interval; as $\lambda$ increases, let us track down the component C defined in Proposition (1.1). First the limit $u_o(\lambda)$ is $C^\infty$ and satisfies (2.2); but its maximum at x = 0 increases monotonically. For some $\lambda_{cr}$, the maximum of $u_0(\lambda)$ is exactly $+A$. As $\lambda > \lambda_{cr}$, an interval $[-a(\lambda), +a(\lambda)]$ appears where $u_0(\lambda) \equiv +A$.

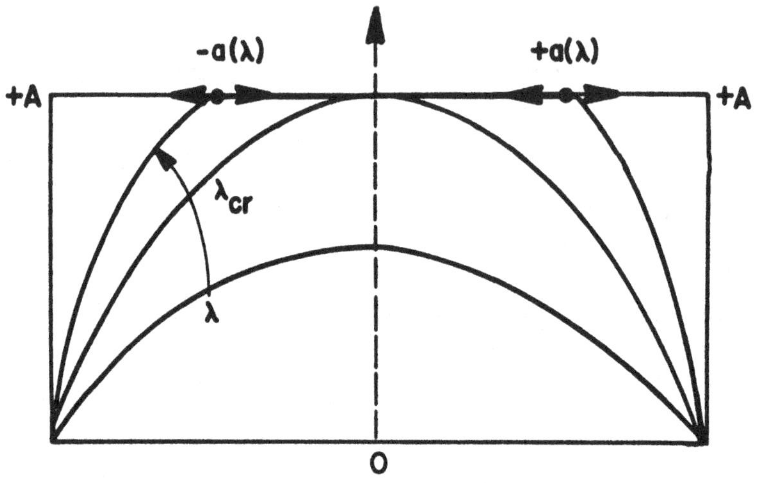


Figure 12

In general, we have the

Proposition 2.4. Let $u_\varepsilon(\lambda)$ a maximal or a minimal solutions of $P_\varepsilon = 0$; then we can extract a subsequence which converges in $H_0^1(\Omega)$ weak to a solution of the F.P.B.

$$\begin{cases} \Delta u_0 + \lambda e^{u_0} \geq 0 \text{ in } \Omega \\ A - u_0 \geq 0 \text{ in } \Omega \\ (A - u_0)(\Delta u_0 + \lambda e^{u_0}) = 0 \text{ in } \Omega \\ u_0/\partial\Omega = 0 \ . \end{cases} \tag{2.21}$$

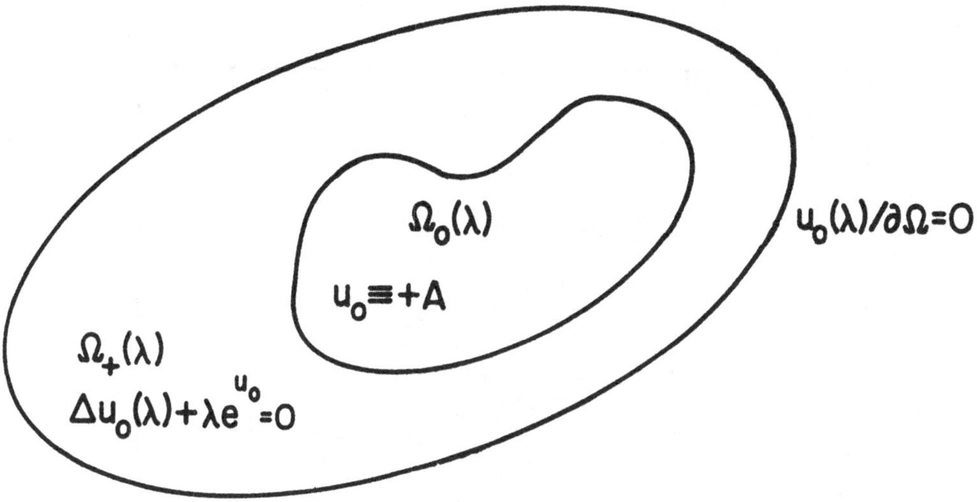


Figure 13

Remark. (2.21) is equivalent to the variational inequality:

$$(\nabla u_0(\lambda) \quad , \nabla(w - u_0(\lambda)) + (\lambda e^{u_0} \quad , w - u_0(\lambda)) \geq 0 \quad , \tag{2.22}$$

$$\forall w \in K = \{w \in H_0^1(\Omega) | w \leq +A\} \quad \text{and} \quad 0 \leq u_0 \leq A \quad .$$

But the remarkable feature is that the singular limit F.B.P. (2.21)-(2.22) itself exhibits multiple solutions and bending points!

Figure 14 (resp. 15) plots global solutions of the F.B.P. (2.22)-(2.23) in a two-dimensional ball (resp. three-dimensional), for various increasing values of the "obstacle" parameter A. Tracking the connected component starting at $\lambda = 0$, $u = 0$, we see that it is first

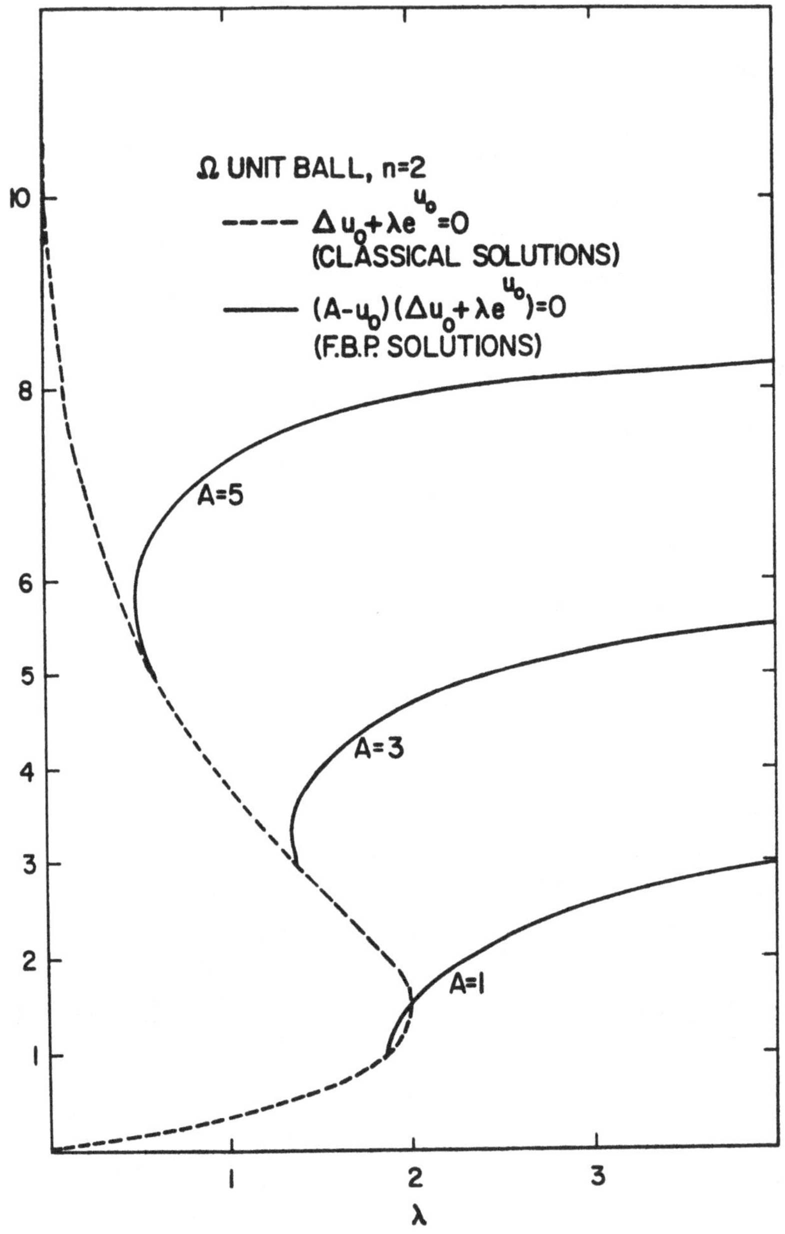
Ω UNIT BALL, n=2
$\Delta u_0 + \lambda e^{u_0} = 0$
(CLASSICAL SOLUTIONS)
$(A-u_0)(\Delta u_0 + \lambda e^{u_0}) = 0$
(F.B.P. SOLUTIONS)
A=5
A=3
A=1
10
8
6
5
4
3
2
1
1
2
3
λ

Figure 14

identical to the classical solution of Gelfand's problem. Then at some $\lambda_{cr}$, a genuine free boundary solution bifurcates from the connected component. In Fig. 14, we see that such a bifurcation (transition) point can occur either before or after the first bending point of Gelfand's problem. Moreover, for A large enough (cf. A = 5, A = 7 curves), the F.B.P. curve itself exhibits a bending point!

In Fig. 15 (n = 3), the Gelfand's problems curve itself admits an infinite number of bending points. We see that the bifurcation (transition) point can occur at or betwen any bending points of Gelfand's curve, following increasing values of A. Also, as A increases, the F.B.P. branch exhibits a bending point of its own, which seems to coalesce to $\lambda = 0$.

The question of which branch of $P_\varepsilon(\lambda,u)$ converges to which branch of the F.B.P. is still a major open one.

In his thesis [33], CONRAD has developed a differential calculus for transition points between classical solutions - F.B.P. solutions.

## 2.5 A N.L.E.P. Whose Singular Limit is a F.B.P. with an Infinity of Solutions

We now give an example where $P_\varepsilon(\lambda,u)$ is multivalued, $P_o(\lambda,u)$ has an infinite number of solutions, and the true singular limit F.B.P. has also an infinite number of solutions. This was investigated by

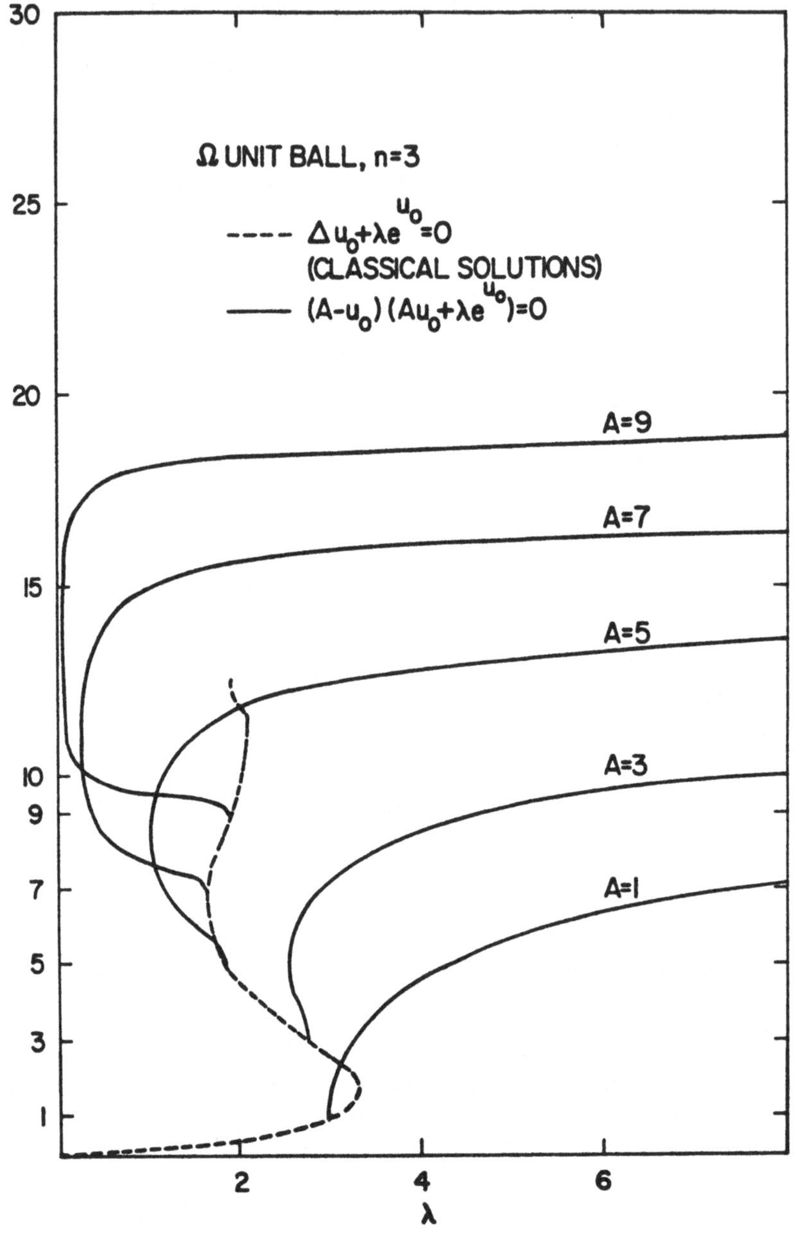
Ω UNIT BALL, n=3
$\Delta u_0+\lambda e^{u_0}=0$
(CLASSICAL SOLUTIONS)
$(A-u_0)(Au_0+\lambda e^{u_0})=0$
A=9
A=7
A=5
A=3
A=1
30
25
20
15
10
9
7
5
3
1
2
4
6
λ

Figure 15

the authors in [6-7,34]:

$$P_\varepsilon(\lambda,u)\begin{cases}\Delta u_\varepsilon = \lambda\dfrac{u_\varepsilon^{m}}{\varepsilon+u_\varepsilon^{m+k}} , \\ u/\partial\Omega = 1 \\ m \geq 1 \quad , 0 < k < 1 \quad ;\end{cases} \tag{2.23}$$

this is the same problem as in (1.3)-(1.4); summarizing the results (cf. Fig. 4):

Proposition 2.5.

$\forall\ \varepsilon > 0$ fixed, $\forall\ \lambda \geq 0$, $P_\varepsilon$ has at least one solution; each solution verifies

$$0 < u_\varepsilon(\lambda) \leq 1 \quad ;$$

as $\lambda \to \infty$, $\varepsilon$ fixed, $u_\varepsilon \to 0$ in $L^P(\Omega)$, with stronger convergence results given by proposition (1.4).

The formal limit problem is given by:

$$P_0(\lambda,u)\begin{cases}\Delta u = \dfrac{\lambda}{u^k} , \\ u/\partial\Omega = 1 \quad , 0 < k < 1 \quad ,\end{cases} \tag{2.24}$$

which was investigated in [6]:

Proposition 2.6.

There exist two cut-off dimensions $n_-(k)$, $n_+(k)$, such that for $n_-(k) < n < n_+(k)$, $P_0(\lambda,u)$ in (2.24) has an infinite number of solutions in n-dimensional balls for some critical $\lambda_c(n,k)$; there are no classical solutions beyond some $\lambda^*$. As $\lambda \to \lambda_c$, min $u_0 \to 0$.

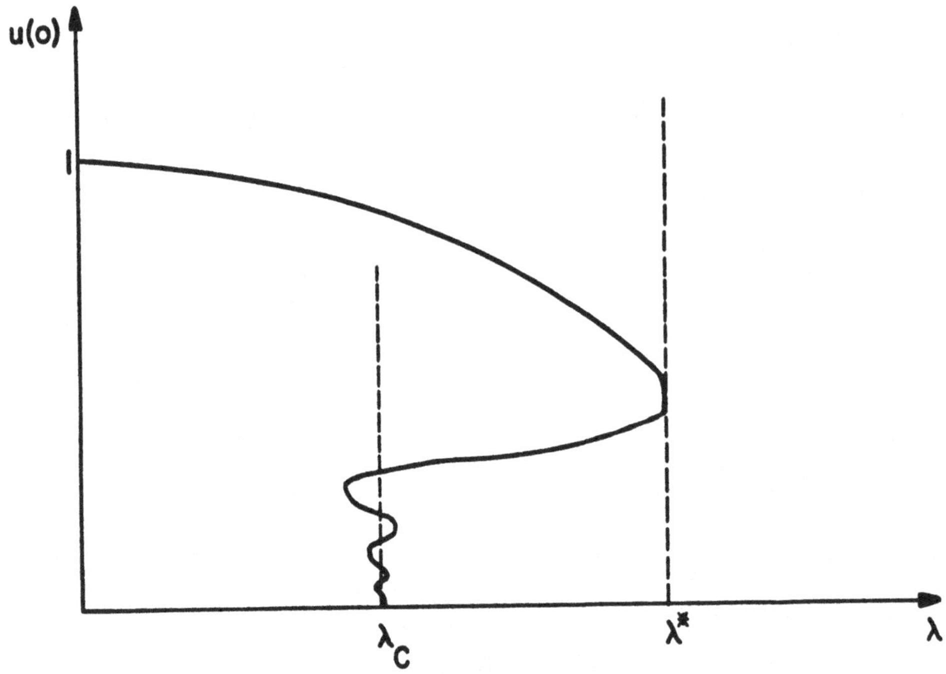


Figure 16

Now as $\varepsilon \to 0^+$, $u_\varepsilon(\lambda)$ converges, for $\lambda$ fixed, to a solution of the following F.B.P. which is not characterized by a Variational Inequality formulation:

Proposition 2.7.

As $\varepsilon \to 0^+$, $u_\varepsilon(\lambda) \to u_o(\lambda)$ in $H^1(\Omega)$ weak, where $u_o(\lambda)$ satisfies:

$$\begin{cases} -\Delta u_o + \dfrac{\lambda}{u_o^k} = 0 \text{ in some } \Omega_+(\lambda) \subset \Omega \quad ; \\ u_o \equiv 0 \text{ in } \Omega_o(\lambda) = \Omega \backslash \Omega_+(\lambda) \quad ; \\ u_o/\partial\Omega = 1 \quad , \end{cases} \tag{2.25}$$

and for extremal solutions:

$$u_o^{-k} \in L^1(\Omega)$$

The situation is summarized in Figure 17, where $\Omega$ is a unit ball, and $R_c(\lambda)$ is the radius of $\Omega_o(\lambda)$.

Remark: For $\lambda$ small enough, $\Omega_o \equiv \emptyset$, and solutions of (2.25) are classical solutions of (2.24). But for $\lambda > \lambda^*$, all solutions of (2.25) have non-trivial free boundaries.

We can now consider (2.25) as a global F.B.P. and study all its solutions, irrespectively of whether they are limits of (2.23) or not. For unit balls such that $n_-(k) < n < n_+(k)$, the results are summarized in Figure 18. Again, $R_c(\lambda)$ denotes the radius of $\Omega_o(\lambda)$ for a nontrivial free boundary solution. The salient features are:

i) the F.B.P. has an infinite number of non-trivial free boundary solutions for the same $\lambda_c$ as $P_o(\lambda,u)$; in

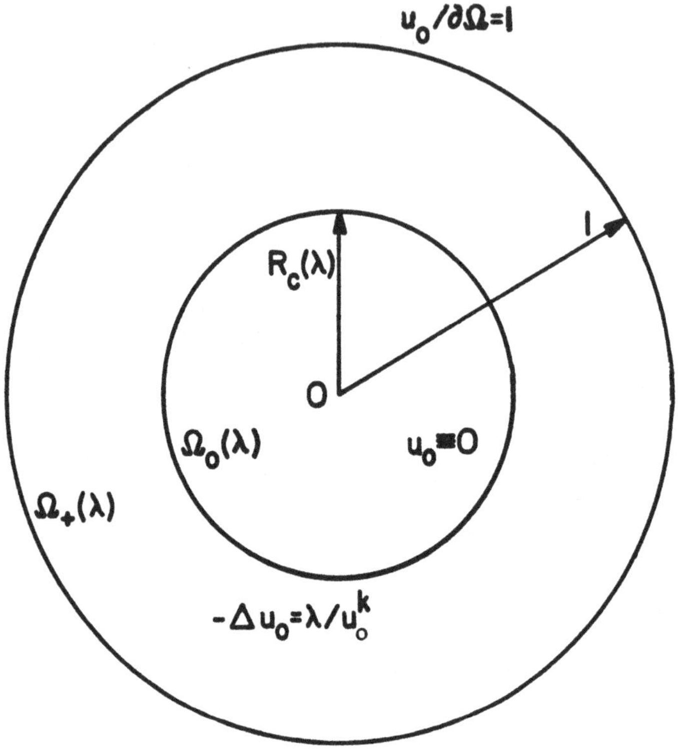


Figure 17

particular, it exhibits an infinite number of bending points; this is truly a <u>nonlinear eigenvalue F.B.P</u>.

ii) the non-trivial free boundary solutions bifurcate away from the classical solutions branch at the angular point $(\lambda_c, u_c)$; at this point, $\Omega_0 = \{0\}$, and the Laplacian blows-up.

In fact, as Figure 18 shows, the global picture for non-trivial free boundary solutions is literally a mirror image of the one for classical solutions (2.24).

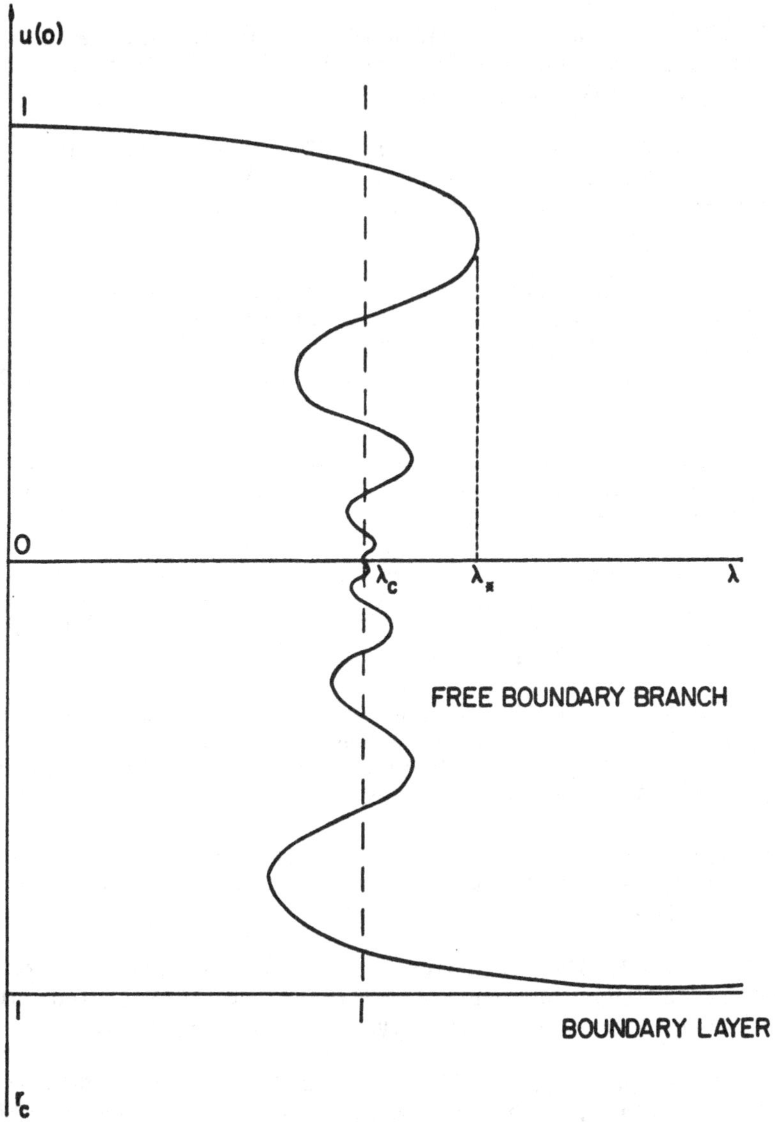


Figure 18

## Bibliography

[1] M. G. CRANDALL and P. H. RABINOWITZ, Bifurcation from simple eigenvalues, J. Funct. Anal., 8, p. 321-340 (1971).

[2] M. G. CRANDALL and P. H. RABINOWITZ, Bifurcation, perturbation of simple eigenvalues and linearized stability, Arch. Rat. Mech. Anal., 52, p. 161-180 (1973).

[3] M. G. CRANDALL and P. H. RABINOWITZ, Some continuation and variational methods for positive solutions of nonlinear elliptic eigenvalue problems, Arch. Rat. Mech. Anal., 58, p. 207-218 (1975).

[4] P. H. RABINOWITZ, Some global results for nonlinear eigenvalue problems, J. Funct. Anal. 7, p. 487-513 (1971). See also: H. AMANN, Fixed point equations and nonlinear eigenvalue problems in ordered Banach spaces, SIAM Review 18, p. 620-709 (1976).

[5] C. M. BRAUNER and B. NICOLAENKO, Sur une classe de problèmes elliptiques non linéaires, C. R. Acad. Sc. Paris, Série A, 286, p. 1007-1010 (1978).

[6] C. M. Brauner and B. Nicolaenko, Sur des problèmes aux valeurs propres non linéaires qui se prolongent en problèmes à frontière libre, C. R. Adac. Sc. Paris, Série A, 287, p. 1105-1100 (1978), and 288, p. 125-127 (1979).

[7] C. M. BRAUNER and B. NICOLAENKO, On nonlinear eigenvalue problems which extend into free boundaries problems, Proc. Conf. on Nonlinear Eigenvalue Problems, p. 61-100, Lect. Notes in Math., 782, Springer-Verlag (1979).

[8] N. I. BRISH, On boundary value problems for the equation $\varepsilon y'' = f(x,y,y')$ for small $\varepsilon$, Dokl. Akad. Nank SSSR 95, p. 429-432 (1954).

[9] Yu. I. BOGLAEV, The two-point problem for a class of ordinary differential equations with a small parameter coefficient of the derivative, USSR Comp. Math. Phys. 10, p. 191-204 (1970).

[10] P. C. FIFE, Semilinear elliptic boundary value problems with small parameters, Arch. Rat. Mech. Anal. 52, p. 205-232 (1973).

[11] J. M. DE VILLIERS, A uniform asymptotic expansion of the positive solution of a nonlinear Dirichlet problem, Proc. London Math. Soc. 27, p. 701-722 (1973).

[12] M. S. BERGER and L. E. FRAENKEL, On the asymptotic solution of a nonlinear Dirichlet problem, J. Math. Mech. 19, p. 553-585 (1970).

[13] A. VAN HARTEN, Nonlinear singular perturbation problems: Proofs of correctness of a formal approximation based on contraction principle in a Banach space, J. Math. Anal. Appl. 65, p. 126-168 (1978).

[14] F. A. HOWES, Boundary-interior layer interactions in nonlinear singular perturbation theory, Memoirs of the A.M.S. 203 (1978).

[15] W. ECKHAUS, Asymptotic analysis of singular perturbations, Studies in Math. 9, North Holland (1979).

[16] H. B. VASILEVA, Asymptotic behavior of solutions to certain problems involving nonlinear differential equations containing a small parameter multiplying the highest derivatives, Russian Math. Surveys 18, p. 13-84 (1963).

[17] R. E. O'MALLEY, Phase-plane solutions to some singular perturbation problems, J. Math. Anal. Appl. 54, p. 149-466 (1976).

[18] Ph. CLEMENT and L. A. PELETIER, On positive and concave solutions of two point nonlinear eigenvalue problems, J. Math. Anal. Appl. 69, p. 329-340 (1979).

[19] Ph. CLEMENT and L. A. PELETIER, Sur les solutions superharmoniques de problèmes aux valeurs propres elliptiques, Proc. Acad. Sc. Paris, to appear (1982).

[20] Ph. CLEMENT and L. A. PELETIER, On positive superharmonic solutions to semi-linear elliptic eigenvalue problems, to appear (1982).

[21] P. C. FIFE, Transition layers in singular perturbation problems, J. Diff. Eqs. 15, p. 77-105 (1974).

[22] P. C. FIFE, Boundary and interior transition layer phenomena for pairs of second-order differential equations, J. Math. Anal. Appl. 54, p. 497-521 (1976).

[23] P. C. FIFE and W. M. GREENLEE, Interior transition layers for elliptic boundary value problems with a small parameter, Uspecki. Matem. Nauk. SSSR 24, p. 103-130 (1974) [Russian Math. Survey 29, p. 103-131 (1975).]

[24] R. E. O'MALLEY, Introduction to singular perturbations, Academic Press (1974).

[25] S. HABER and N. LEVINSON, A boundary value problem for a singularly perturbed differential equation, Proc. Amer. Math. Soc. 6, p. 866-872 (1955).

[26] F. A. HOWES, A class of boundary value problems whose solutions possess angular limiting behavior, Rocky Mtn. J. Math. 6, p. 591-607 (1976).

[27] R. E. O'MALLEY, On nonlinear singular perturbation problems with interior nonuniformities, J. Math. Mech. 19, p. 1103-1112 (1970).

[28] C. M. BRAUNER and B. NICOLAENKO, Perturbation singulière, solutions multiples et hystérésis dans un problème de biochimie, C. R. Acad. Sc. Paris, Série A, 283, p. 775-778 (1976).

[29] C. M. BRAUNER and B. NICOLAENKO, Singular perturbation, multiple solutions and hystérésis in a nonlinear problem, Lect. Notes in Math., 594, Springer-Verlag, p. 50-76 (1977).

[30] C. M. BRAUNER and B. NICOLAENKO, Singular perturbations and free boundary value problems, Proc. 4th Int. Symp. on Computing methods in Applied Sciences and Engineering, p. 669-724, North Holland (1980).

[31] G. DUVAUT and J. L. LIONS, Les inequations en Mécanique et Physique, Dunod, Paris (1972).

[32] D. KINDERLEHRER and G. STAMPACCHIA, Introduction to Variational inequalities and Application, Academic Press (1979).

[33] F. CONRAD, Doctoral Thesis, Ecole Centrale de Lyon and Universite de Saint-Etienne, France (1982).

[34] C. M. BRAUNER and B. NICOLAENKO, Free boundary value problems as singular limits of nonlinear eigenvalue problems, Proc. Symp. on Free Boundary Problems, Pavia, E. Magenes Ed., Istituto Nazionale di Alta Matematica Francesco Severi (Roma) Publisher, Vol. II, p. 61-84 (1980).

# SINGULARLY PERTURBED SYSTEMS OF DIFFUSION TYPE AND FEEDBACK CONTROL

A. van Harten
Mathematical Institute
Un. of Utrecht
The Netherlands

## 1. Introduction

In this paper we shall consider linear systems of diffusion-type subject to a certain feed-back control mechanism in a situation, where the diffusion constant is a small parameter. Such controlled diffusion systems can be found f.e. in the context of heating problems [1], [2] or chemical or nuclear reactor design, [3]. For the feedback control there are many possibilities: feedback without or with memory, with distributed input or input through the boundary, etc., while it also depends on the number and kind of observations, cf. [2], [4], [5], [17]. Here we shall consider distributed as well as boundary control, but always on the basis of an instantaneous feedback coupling using observations of the state in a finite number of points $y_1,\ldots,y_p$ in the interior of the domain D. In the case of Dirichlet boundary conditions the evolution of the state is described by one of the following problems:

$$(1.1)_d \quad \frac{\partial u}{\partial t} = L_\varepsilon u + \tilde{\Pi}_d u + h \qquad \text{(distributed control)}$$

$$u = s \text{ on } \partial D$$

$$u(.,0) = \psi$$

$$(1.1)_b \quad \frac{\partial u}{\partial t} = L_\varepsilon u + h \qquad \text{(boundary control)}$$

$$u = \tilde{\Pi}_b u + s \text{ on } \partial D$$

$$u(.,0) = \psi$$

Here $L_\varepsilon$, $\tilde{\Pi}_d$, $\tilde{\Pi}_b$ are of the following form

$$(1.2) \quad L_\varepsilon = \varepsilon L_2 - L_r$$

with $L_2 = \sum_{i,j=1}^{n} \frac{\partial}{\partial x_i} a_{ij} \frac{\partial}{\partial x_j}$, $L_r = \sum_{i=1}^{n} v_i \frac{\partial}{\partial x_i} + \gamma$

$$(1.3)_d \quad \tilde{\Pi}_d u = c_0 + \sum_{i=1}^{p} c_i(\delta_{y_i} u - I_i) \qquad c_0, c_i \in C^\infty(\overline{D})$$

$$(1.3)_b \quad \tilde{\Pi}_b u = b_0 + \sum_{i=1}^{p} b_i(\delta_{y_i} u - I_i) \qquad b_0, b_i \in C^\infty(\partial D)$$

We suppose, that $\overline{D} \subset \mathbb{R}^n$ is a compact set with a smooth boundary $\partial D$. The coefficients $a_{ij}$, $v_i$, $\gamma$ are also supposed to be smooth. Further we assume, that $L_2$ is uniformly elliptic. By $\delta_{y_i}$ we denote the continuous linear functional on $C(\overline{D})$, which maps $u \to u(y_i)$. Note, that the feedback control consists of a part independent of the observations $\delta_{y_i} u$ and a part proportional to the difference between the $\delta_{y_i} u$'s and certain ideal values $I_i$. We shall always assume that the observation points $y_i$ have an $O(1)$ distance to the boundary $\partial D$.

Because of the small paramter $\varepsilon$ in front of the highest order derivatives, the problems $(1.1)_{b,d}$ have a singular perturbation character. The stationary, uncontrolled problem corresponding to $(1.1)_{b,d}$ has been thoroughly analyzed, [6], [7], [8], [9], [10]. It was understood, that for the behaviour of the solution for $\varepsilon \downarrow 0$ it makes a big difference, whether there is convection: $v \not\equiv 0$, or not: $v \equiv 0$. If there is convection the structure of the velocity field plays an important role, especially the presence of turning points, cycles or tangency points at the boundary. If $v \equiv 0$ the sign of the coefficient $\gamma$ is very relevant. Here we shall consider the following two cases:

$(1.4)_0$ $\quad r = 0: L_0 u = \gamma u, \qquad \gamma > 0$

and a domain as in fig. 1

$(1.4)_1$ $\quad r = 1: L_1 u = v.\nabla u + \gamma u, \qquad \|v\| > 0$

with a domain and velocity field as in fig. 1.

fig. 1.

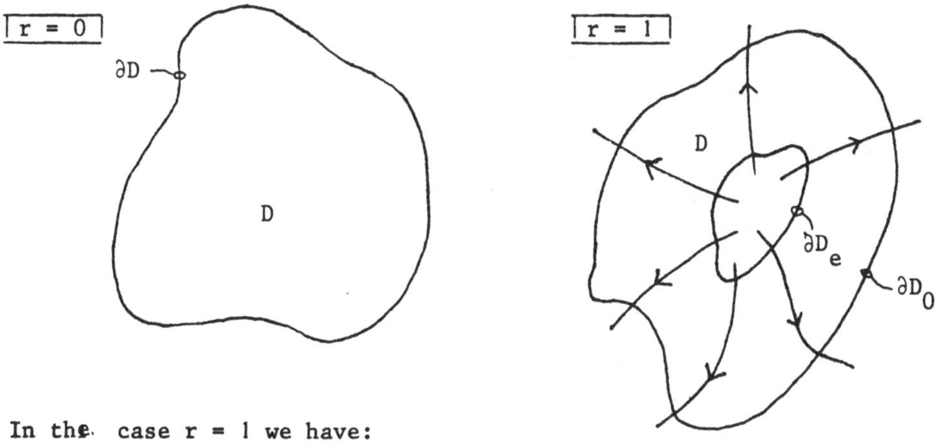


In the case r = 1 we have:

(1.6) $\quad v.n < 0$ on $\partial D_e$; $v.n > 0$ on $\partial D_0$

$n$ = the outer normal on $\partial D$

Further, let $z(t;x)$ parametrize the characteristic through $x$ i.e.

(1.7) $$\frac{dz}{dt} = v(z); \; z(0;x) = x$$

Then, we assume that $\forall x \in \overline{D} \; \exists t_e(x) \leqslant 0$ such that

(1.8) $$z(t_e(x);x) \stackrel{\text{def}}{=} x_e \in \partial D_e$$

Note, that the conditions on the velocity field in $(1.4)_1$-(1.8) are such that turning points or cycles are excluded and at each point of $\partial D$ the field is transverse. Further, using the theory of O.D.E. it is clear, that $t_e(x)$ and $x_e$ are uniquely defined, smooth functions of x.

As for the behaviour for $\varepsilon \downarrow 0$ of the solution of the dynamic, uncontrolled singular perturbation problem corresponding to $(1.1)_{b,d}$, there are only a few references [20], [11], [1]. The asymptotic theory for solutions of such parabolic problems is somewhat less developed than for elliptic problems. In this respect section 3 contains some new contributions for the case of a 1st order unperturbed operator $L_1$ as in $(1.4)_1$.

When now, for a moment, we forget the point of view of asymptotics and take $\varepsilon = \varepsilon_0$ = fixed, there are a large number of results from infinite dimensional control theory, which are applicable. They concern f.e. the following subjects: well-posedness of the controlled problem, generation of semi-groups by the controlled diffusion operator and stabilizability of the system, [2], [4], [12], [13], [14], [15], [17]. Some of these results will be useful in the sequel and sometimes it will be nice to compare our results found by asymptotic calculations with predictions valid for the general case, see for example the topic of stabilizability, section 4.

Our purpose is to use singular perturbation techniques to analyze the behaviour of the solution of the controlled problem $(1.1)_{d,b}$ asymptotically for $\varepsilon \downarrow 0$. In section 2 this is done for the corresponding stationary problem and in section 3 for the dynamic problem. As a result we obtain explicit formal asymptotic approximations as $\varepsilon \downarrow 0$ for the effect of the control and also for the spectrum of the controlled operator. In section 4 it is sketched, why the results of the previous sections are rigorous. In section 5 we use the results found before to construct a near optimal control with respect to a certain cost-functional. Finally, we remark that asymptotic methods are used more often in control theory, but for problems where the small parameter $\varepsilon$ enters in a different way. For example, the asymptotics as considered here, is from a completely different type, than in [16], where the small parameter is in front of the $\frac{\partial}{\partial t}$ term. Other examples of different asymptotics can be found looking through this volume or in [20].

## 2. The stationary, controlled problem

Without loss of generality we can restrict ourselves to the following problems with homogeneous boundary conditions

$$(2.1)_d \quad (L_\varepsilon + \Pi_d - \lambda)u^\varepsilon + f = 0, \qquad u^\varepsilon = 0 \text{ on } \partial D$$

$$(2.1)_b \quad (L_\varepsilon - \lambda)u^\varepsilon + f = 0, \qquad u^\varepsilon = \Pi_b u^\varepsilon \text{ on } \partial D$$

with $\Pi_d = \sum_{i=1}^{p} c_i\delta_{y_i}$ , $\Pi_b = \sum_{i=1}^{p} b_i\delta_{y_i}$.

Note, that we have introduced a spectral parameter $\lambda$ in $(2.1)_{d,b}$. The trick to solve these problems is wellknown from Weinstein-Aronszajn's theory, [18]. If $\lambda \notin \sigma(L_\varepsilon)$ we can rewrite $(2.1)_{d,b}$ as follows

$$(2.2)_d \quad u^\varepsilon = -F^\varepsilon - \sum_{i=1}^{p} \xi_i^\varepsilon C_i^\varepsilon \qquad (2.2)_b \quad u^\varepsilon = -F^\varepsilon + \sum_{i=1}^{p} \xi_i^\varepsilon B_i^\varepsilon$$

with $\xi_i^\varepsilon = \delta_{y_i}u^\varepsilon$, $F^\varepsilon = (L_\varepsilon - \lambda)^{-1}f$, $C_i^\varepsilon = (L_\varepsilon - \lambda)^{-1}c_i^\varepsilon$ and $B_i^\varepsilon$ is the solution of $(L_\varepsilon - \lambda)B_i^\varepsilon = 0$, $B_i^\varepsilon = b_i$ on $\partial D$. Substitution of $x = y_k$ in (2.2) provides us with a linear system of equations for $\xi^\varepsilon \in \mathbb{C}^p$.

$$(2.3)_d \quad [I + \Omega_d^\varepsilon]\xi^\varepsilon = -\eta^\varepsilon \qquad (2.3)_b \quad [I - \Omega_b^\varepsilon]\xi^\varepsilon = -\eta^\varepsilon$$

with: $\eta_k^\varepsilon = \delta_{y_k}F^\varepsilon$, $[\Omega_d^\varepsilon]_{k,i} = \delta_{y_k}C_i^\varepsilon$, $[\Omega_b^\varepsilon]_{k,i} = \delta_{y_k}B_i^\varepsilon$

If $I + \Omega_d^\varepsilon$, $I - \Omega_b^\varepsilon$ are invertible, we only have to put the solution $\xi^\varepsilon$ of $(2.3)_{d,b}$ into $(2.2)_{d,b}$ and the solution $u^\varepsilon$ of $(2.1)_{d,b}$ is known. Note, that then the effect of the control is given by the following expressions

$$(2.4)_d \quad \text{e.d.c.} = \langle (I+\Omega_d^\varepsilon)^{-1}\eta^\varepsilon, C_\cdot^\varepsilon\rangle \qquad (2.4)_b \quad \text{e.b.c.} = -\langle (I-\Omega_b^\varepsilon)^{-1}\eta^\varepsilon, B_\cdot^\varepsilon\rangle$$

with for $e, e' \in \mathbb{C}^p$: $< e,e' > = \sum_{i=1}^{p} e_i e_i'$ and $C^\varepsilon_\cdot$, $B^\varepsilon_\cdot$ the vectors with components $C^\varepsilon_i$, $B^\varepsilon_i$.

Of course the points, where $I + \Omega^\varepsilon_d$, $I - \Omega^\varepsilon_b$ are singular, belong to the spectrum of the controlled operator.

It is now clear, that in order to construct approximations of the solution of the controlled stationary problem for $\varepsilon \downarrow 0$ it is sufficient to have asymptotic approximations for the functions $F^\varepsilon$, $C^\varepsilon_i$ and $B^\varepsilon_i$. These functions are determined by uncontrolled problems of the following type

$$(2.5)_d \quad (L_\varepsilon - \lambda)C^\varepsilon = c \qquad\qquad (2.5)_b \quad (L_\varepsilon - \lambda)B^\varepsilon = 0$$
$$C^\varepsilon = 0 \text{ on } \partial D \qquad\qquad B^\varepsilon = b \text{ on } \partial D$$

Using the method of matched asymptotic expansions approximations C, B of $C^\varepsilon$ and $B^\varepsilon$ are easily found, [6], [7], [8], [9]. Thus we are lead to approximations $\Omega_{d,b}$ of $\Omega^\varepsilon_{d,b}$, $\eta$ of $\eta^\varepsilon$ and $\xi$ of $\xi^\varepsilon$, where $\xi$ satisfies the approximate version of $(2.3)_{d,b}$, i.e. $(1 + \Omega_d)\xi = -\eta$, $(1 - \Omega_b)\xi = -\eta$ and if $1 + \Omega_d$, $1 - \Omega_b$ are invertible we end up with an approximation u of $u^\varepsilon$.

In *the case of a 0th order unperturbed operator* as in $(1.4)_0$ the approximations consist of a regular expansion in the interior of D and a boundary layer of width $\sqrt{\varepsilon}$ along all of $\partial D$:

$$(2.6)_d \quad C^\varepsilon = \hat{C}_0(x) + \hat{G}_0(\zeta,\phi)H(x) + O(\sqrt{\varepsilon})$$

$$(2.6)_b \quad B^\varepsilon = \overline{G}_0(\zeta,\phi)H(x) + O(\sqrt{\varepsilon})$$

with $\zeta$ = distance to $\partial D/\sqrt{\varepsilon}$, $\phi$ = a (local) parametrization of $\partial D$, H(x) a suitably chosen $C^\infty$ cutoff function. Note, that in the case of $(2.5)_b$ the regular expansion is $\equiv 0$ and the approximation is completely of layer type. The functions $\hat{C}_0$, $\hat{G}_0$ and $\overline{G}_0$ are found as the solutions of the following problems:

$$(2.7) \qquad -(\gamma + \lambda)\hat{C}_0 = c$$

$$(2.8)_d \quad \left(a \frac{d^2}{d\zeta^2} - \bar{\gamma} - \lambda\right)\hat{G}_0 = 0 \qquad (2.8)_b \quad \left(a \frac{d^2}{d\zeta^2} - \bar{\gamma} - \lambda\right)\bar{G}_0 = 0$$

$$\hat{G}_0|_{\zeta=0} = -\hat{C}_0|_{\partial D} \qquad \bar{G}_0|_{\zeta=0} = b$$

$$\lim_{\zeta\to\infty} \hat{G}_0 = 0 \qquad \lim_{\zeta\to\infty} \bar{G}_0 = 0$$

with $a = \Sigma n_i (a_{ij})|_{\partial D} n_j > 0$, $\bar{\gamma} = \gamma|_{\partial D} > 0$. Hence:

$$(2.9)A \quad \hat{C}_0 = -c/(\gamma + \lambda), \quad \hat{G}_0 = -\hat{C}_0|_{\partial D} \exp(-\mu\zeta)$$

$$(2.9)B \quad \bar{G}_0 = b \exp(-\mu\zeta)$$

with $\mu = \sqrt{(\bar{\gamma} + \lambda)/a}$. In order to be able to divide by $\gamma + \lambda$ and to have exponential decay of the boundary layer terms we must have

$$(2.10) \qquad \lambda \notin (-\infty, -\hat{\gamma}] \text{ with } \hat{\gamma} = \min_{x\in\bar{D}} \gamma(x) > 0$$

Now using the approximations as found in (2.6-9) we find:

$$(2.11)_d \quad \Omega_d^\varepsilon = \Omega_d + O(\sqrt{\varepsilon}), \qquad (2.11)_b \quad \Omega_b^\varepsilon = \Omega_b + O(\sqrt{\varepsilon})$$

$$\text{with } [\Omega_d]_{k,i} = -\frac{c_i(y_k)}{\lambda + \gamma(y_k)}, \qquad \Omega_b = 0.$$

The conclusion is, that in the case of boundary control (i) the effect of the control is only noticable in a layer of width $\sqrt{\varepsilon}$ along $\partial D$ and (ii) the spectrum of the controlled operator is contained in a set, which shrinks with $\varepsilon \downarrow 0$ to $(-\infty, -\hat{\gamma}]$. For distributed control the spectrum is contained in a set, which shrinks with $\varepsilon \downarrow 0$ to $(-\infty, -\hat{\gamma}] \cup \{\lambda_1, \ldots, \lambda_q\}$ with $q \leq p$ and $\lambda_1, \ldots, \lambda_q$ the eigenvalues of the matrix $\Lambda$, where

$$(2.12)_d \quad \Lambda_{k,i} = -\gamma(y_k) + c_i(y_k)$$

More precise information on the set, which contains the spectrum is given in section 4.

Let us now consider *the case of a 1st order unperturbed operator* as in $(1.4)_1$. Then the approximations of the solutions of $(2.5)_{d,b}$ have a different structure. They consist of a regular expansion valid upto $\partial D_e$ and a boundary layer of width $\varepsilon$, along $\partial D_0$.

In order to describe these approximations it is simpler to introduce the following coordinates.

(2.13) $s = -t_e(x);\quad \phi = x_e$

Note, that in these coördinates $\partial D_e = \{s = 0\}$ and $v.\nabla = \frac{\partial}{\partial s}$. The other part of the boundary, $\partial D_0$, can be given as $\{(s,\phi) \mid s = T(\phi)\}$, where the interpretation of $T(\phi)$ is: the time it takes to travel along the characteristic through $(0,\phi)$ from $(0,\phi)$ to $\partial D_e$. Now our approximation will have the following form

$$(2.14)_d \quad C^\varepsilon = \hat{C}_0(s,\phi) + \hat{G}_0^0(\zeta,\phi)H(T(\phi) - s) + O(\varepsilon)$$

$$(2.14)_b \quad B^\varepsilon = \overline{B}_0(s,\phi) + \overline{G}_0^0(\zeta,\phi)H(T(\phi) - s) + O(\varepsilon)$$

with $\zeta = (T(\phi) - s)/\varepsilon$ and H a suitably chosen $C^\infty$ cutoff function. The functions $\hat{C}_0$, $\hat{G}_0$, $\overline{B}_0^0$, $\overline{G}_0^0$ are found as the solutions of:

$$(2.15)_d \quad (-\frac{\partial}{\partial s} - \gamma - \lambda)\hat{C}_0 = c \qquad (2.15)_b \quad (-\frac{\partial}{\partial s} - \gamma - \lambda)\overline{B}_0 = 0$$

$$\hat{C}_0(0,\phi) = 0 \qquad \overline{B}_0(0,\phi) = b_e(\phi)$$

$$(2.16)_d \quad (A^0 \frac{\partial^2}{\partial \zeta^2} + \frac{\partial}{\partial \zeta})\hat{G}_0^0 = 0 \qquad (2.16)_b \quad (A^0 \frac{\partial^2}{\partial \zeta^2} + \frac{\partial}{\partial \zeta})\overline{G}_0^0 = 0$$

$$\hat{G}_0^0(0,\phi) = -\hat{C}_0(T(\phi),\phi) \qquad \overline{G}_0^0(0,\phi) = b_0(\phi) - \overline{B}_0(T(\phi),\phi)$$

$$\lim_{\zeta\to\infty} \hat{G}_0^0(\zeta,\phi) = 0 \qquad \lim_{\zeta\to\infty} \overline{G}_0^0(\zeta,\phi) = 0$$

with $A^0 = \Sigma(\frac{\partial t_e}{\partial x_i} a_{ij} \frac{\partial t_e}{\partial x_i})\Big|_{\partial D_0}$; $b_e$ and $b_0$ are the values of b on $\partial D_e$ and $\partial D_0$, respectively.

It is easy to check, that $\hat{C}_0$, $\hat{G}_0^0$, $\overline{B}_0$ and $\overline{G}_0^0$ are given by

$$(2.17) \quad \hat{C}_0(s,\phi) = -\int_0^s c(\overline{s},\phi) \exp\,[-\int_{\overline{s}}^s \gamma(s',\phi)ds' - \lambda(s - \overline{s})]\, d\overline{s}$$

$$\hat{G}_0^0(\zeta,\phi) = -\hat{C}_0(T(\phi),\phi) .\ \exp(-\zeta/A^0)$$

$$(2.17)_b \quad \overline{B}_0(s,\phi) = b_e(\phi) .\ \exp\,[-\int_0^s \gamma(s',\phi)ds' - \lambda s]$$

$$\overline{G}_0^0(\zeta,\phi) = [b_0(\phi) - \overline{B}_0(T(\phi),\phi)] \ .\ \exp(-\zeta/A^0)$$

Now, using the approximations as found in (2.14-17) we obtain:

$$(2.18)_d \quad \Omega_d^\varepsilon = \Omega_d(\lambda) + O(\varepsilon)$$

with $[\Omega_d(\lambda)]_{k,i} = -\int_0^{s_k} c_i(\bar{s},\phi_k)\exp[-\int_{\bar{s}}^{s_k} \gamma(s',\phi_k)ds' - \lambda(s_k - \bar{s})]d\bar{s}$

where $(s_k,\phi_k)$ denotes the point $y_k$ in $(s,\phi)$ coordinates, i.e. $s_k =$ $= -t_e(y_k)$; $\phi_k = (y_k)_e$.

$$(2.18)_b \quad \Omega_b = \Omega_b(\lambda) + O(\varepsilon)$$

with $[\Omega_b(\lambda)]_{k,i} = b_{i,e}(\phi_k)\exp[-\int_0^{s_k} \gamma(s',\phi_k)ds' - \lambda s_k]$

Hence, the spectrum of the controlled operator is contained in a set, which shrinks to the zero's of a holomorphic function $\omega_d(\lambda)$, $\omega_b(\lambda)$ in the respective cases of distributed control and boundary control with:

$$(2.19)_d \quad \omega_d(\lambda) = \det[I + \Omega_d(\lambda)], \qquad (2.19)_b \quad \omega_b(\lambda) = \det[I - \Omega_b(\lambda)]$$

Of course, it would be interesting to have a rough idea about the location of the zero's of these holomorphic functions. Using integration by parts it is not difficult to show that in the case of distributed control $\forall A \in \mathbb{R} \quad \exists B > 0$ such that for all $\lambda$ with Re $\lambda \geqslant A$: $[\Omega_d(\lambda)]_{i,j} \leqslant$ $\leqslant \frac{B}{1 + |\lambda|}$. Then an application of Gershgorin's theorem, [19], shows, that the zero's $\{\lambda_k;\ k \in \mathbb{N}\}$ of $\omega_d(\lambda)$ can be numbered in such a way that Re $\lambda_k \downarrow -\infty$ for $k \uparrow \infty$. However, in the case of boundary control the situation is quite different. It is easy to verify, that

$$\begin{aligned}\omega_b(\lambda) &= \exp(-(\lambda - \lambda_0)\mathrm{tr}(S)).\det[e^{(\lambda-\lambda_0)S} - \Omega_b(\lambda_0)] \\ &= \exp(-(\lambda - \lambda_0)\mathrm{tr}(S)).\det[e^{(\lambda-\lambda_0)S}\Omega_b^{-1}(\lambda_0) - I]\det[\Omega_b(\lambda_0)]\end{aligned}$$

Here S denotes the diagonal matrix with $S_{k,k} = s_k$ and the only requirement for $\lambda_0$ is, that $\Omega_b(\lambda_0)$ is non-singular. Using again Gershgorin's theorem we see, that the zero's of $\omega_b(\lambda)$ lie in a strip $\{\lambda \mid \alpha < \mathrm{Re}\ \lambda < \beta\}$.

More detailed information on the location of the spectrum of the controlled operator can again be found in section 4. As for the effect of the control we observe, that in the case of boundary control the control input on $\partial D_0$ is only noticable in the boundary layer of width $\varepsilon$ along $\partial D_0$.

On the basis of the results we derived for the asymptotic location of the spectrum of the controlled operator, we expect that an approximation $\bar{u}$ of the solution of the dynamic problem (see $(3.1)_{d,b}$), will grow not faster than

(2.20) $\quad |\bar{u}(x,t)| \leq C(\nu,\varepsilon)e^{\nu t}$

In this estimate we can presumably take $\nu \in \mathbb{R}$ and $\nu > \nu_d^r$ in the case of distributed control with $\nu_d^0 = \max(-\hat{\gamma}, \mathrm{Re}\,\lambda_1,\ldots,\mathrm{Re}\,\lambda_q)$, $\nu_d^1 =$ $= \max\{\mathrm{Re}\,\lambda | \omega_d(\lambda) = 0\}$ and in the case of boundary control $\nu \in \mathbb{R}$, $\nu > \nu_b^r$ with $\nu_b^0 = -\hat{\gamma}$, $\nu_b^1 = \sup\{\mathrm{Re}\,\lambda \,|\omega_b(\lambda) = 0\}$.

In the next section we shall see, that an estimate as in (2.20) indeed holds and in addition we shall find how the constant $C(\nu,\varepsilon)$ in (2.20) depends on $\varepsilon$.

## 3. The dynamic, controlled problem

Here we shall consider the time evolution of the state, when the equation and the boundary conditions are homogeneous:

$(3.1)_d$
$$\frac{\partial \bar{u}^\varepsilon}{\partial t} = (L_\varepsilon + \Pi_d)\bar{u}^\varepsilon$$
$$\bar{u}^\varepsilon = 0 \text{ on } \partial D$$
$$\bar{u}^\varepsilon(.,0) = \psi \in C^\infty(\bar{D})$$
$$\psi = 0 \text{ on } \partial D$$

$(3.1)_b)$
$$\frac{\partial \bar{u}^\varepsilon}{\partial t} = L_\varepsilon \bar{u}^\varepsilon$$
$$\bar{u}^\varepsilon = \Pi_b \bar{u}^\varepsilon \text{ on } \partial D$$
$$\bar{u}^\varepsilon(.,0) = \psi \in C^\infty(\bar{D})$$
$$\psi = \Pi_b \psi \text{ on } \partial D$$

In order to solve these problems we denote the observations $\delta_{y_k}\bar{u}^\varepsilon(.,t)$ by $\xi_k^\varepsilon(t)$. The solutions of $(3.1)_{d,b}$ can then be expressed in the following way:

$$(3.2)_d \quad \overline{u}^{\varepsilon}(.,t) = e^{L_\varepsilon t}\psi + \sum_{i=1}^{p}\int_0^t \xi_i^{\varepsilon}(t-\tau)e^{L_\varepsilon \tau}c_i\,d\tau$$

$$(3.2)_b \quad \overline{u}^{\varepsilon}(.,t) = e^{L_\varepsilon t}\psi + \frac{\partial}{\partial t}\{\sum_{i=1}^{p}\int_0^t \xi_i^{\varepsilon}(t-\tau).(1-e^{L_\varepsilon \tau})B_i^{\varepsilon}d\tau\}$$

Here $B_i^\varepsilon$ denotes the solution of the uncontrolled, stationary problem: $L.B_i^\varepsilon = 0$, $b_i^\varepsilon = b_i$ on $\partial D.B_i^\varepsilon$ is well-defined, since $0 \notin \sigma(L_\varepsilon)$, see section 4.

By $v^\varepsilon(.,t) = e^{L_\varepsilon t}\chi$ we denote the solution of the uncontrolled, dynamic problem starting at $t = 0$ in $\chi$:

$$(3.3) \quad \frac{\partial v^\varepsilon}{\partial t} = L_\varepsilon v^\varepsilon$$

$$v^\varepsilon = 0 \text{ on } \partial D$$

$$v^\varepsilon(.,0) = \chi$$

Substitution of $x = y_k$ in $(3.2)_{d,b}$ yields the following Volterra equations for $\xi^\varepsilon(t)$

$$(3.4)_d \quad \xi^\varepsilon(t) = \eta^\varepsilon(t) + \int_0^t K_d^\varepsilon(\tau)\xi^\varepsilon(t-\tau)d\tau$$

$$(3.4)_b \quad \xi^\varepsilon(t) = \eta^\varepsilon(t) + \frac{d}{dt}\{\int_0^t K_b^\varepsilon(\tau)\xi^\varepsilon(t-\tau)d\tau\}$$

with $\eta_k^\varepsilon(t) = \delta_{y_k} e^{L_\varepsilon t}\psi$, $[K_d^\varepsilon(\tau)]_{k,i} = \delta_{y_k} e^{L_\varepsilon \tau} c_i$, $[K_b^\varepsilon(\tau)]_{k,i} = \delta_{y_k}(1-e^{L_\varepsilon \tau})B_i^\varepsilon$. Once the solutions $\xi^\varepsilon(t)$ of these Volterra equations $(3.4)_{d,b}$ are known we find the solutions $\overline{u}^\varepsilon$ of $(3.1)_{d,b}$ simply by substitution of $\xi^\varepsilon(t)$ in $(3.2)_{d,b}$. Let us now consider our task, the construction of asymptotic approximation for $\varepsilon \downarrow 0$ of the solutions of $(3.1)_{d,b}$. This task really boils down to finding an asymptotic approximation $v$ for $\varepsilon \downarrow 0$ of the solution $v^\varepsilon$ of the uncontrolled problem (3.3). In this respect it is important to notice, that an asymptotic approximation of $B_i^\varepsilon$ is already known, see section 2, $(2.5)_b$.

Once such an approximation is available we also have approximations $K_{d,b}$ of $K_{d,b}^\varepsilon$, $\eta$ of $\eta^\varepsilon$ and $\xi$ of $\xi^\varepsilon$, where $\xi$ is found as the solution of the approximate version of $(3.3)_{d,b}$. Next an approximation $\overline{u}$ of $\overline{u}^\varepsilon$ is found by substituting all approximations of the r.h.s. of $(3.2)_{d,b}$.

In *the case of an unperturbed operator of 0th order* as in $(1.4)_0$ the approximation consists of a regular expansion corrected by a boundary layer of width $\sqrt{\varepsilon}$ along all of $\partial D$, just as in the stationary case. However, the various terms in the approximation now satisfy dynamic equations.

$$(3.5) \qquad v^{\varepsilon} = V_0(x,t) + \tilde{P}_0(-).H(x) + O(e^{-\gamma t}\sqrt{\varepsilon})$$

$$(3.6) \qquad \frac{\partial V_0}{\partial t} = -\gamma V_0$$

$$V_0(.,0) = \chi$$

$\tilde{P}_0$ is the zero'th order boundary layer term and H is a suitable $C^{\infty}$ cutoff function. In order to construct $\tilde{P}_0$ we introduce rather special coordinates $(y,\theta)$ near $\partial D$, such that: $\partial D = \{y = 0\}$ and

$$(3.7) \qquad L_2 = \tilde{a}\frac{\partial^2}{\partial y^2} + \sum_{i,j=1}^{n-1} \tilde{a}_{ij}\frac{\partial^2}{\partial\theta_i\partial\theta_j} + \tilde{a}_0\frac{\partial}{\partial y} + \sum_{i=1}^{n-1} \tilde{a}_i\frac{\partial}{\partial\theta_i}$$

This can be done by first introducing $(y,\phi)$-coordinates with y the distance from x to $\partial D$ and $\phi$ the parameters of the point on $\partial D$ nearest to x. Next, we define $\theta_i = \phi_i + g_i(y,\phi)$ with $g_i$ the solution of $\partial g_i/\partial y = -\frac{1}{2}\hat{a}_{0,i}(y,\phi)$, $g_i(0,\phi) = 0$, where $\hat{a}_{0,i}$ denotes the coefficient in front of $\partial^2/\partial y\partial\phi_i$ in $L_2$. The reason for the introduction of these special coordinates is to avoid singularities at $\partial D$, $t = 0$ in $(\frac{\partial}{\partial t} - L_{\varepsilon})\tilde{P}_0$, which are produced by the differentiations $\partial^2/\partial y\partial\phi_i$. For the same reason we want to annihilate the effect of the differentiation $\tilde{a}_0\,\partial/\partial_y$ on $\tilde{P}_0$ and therefore we write

$$(3.8) \qquad \tilde{P}_0(-) = \exp\left(-\tfrac{1}{2}\int_0^y \frac{\tilde{a}_0(s,\theta)}{\tilde{a}(s,\theta)}\,ds\right).P_0(\zeta,\theta,t)$$

with $\zeta = y/\sqrt{\varepsilon}$ and $P_0$ the solution of

$$(3.9) \qquad \frac{\partial P_0}{\partial t} = a\frac{\partial^2 P_0}{\partial\zeta^2} - \bar{\gamma}P_0$$

$$P_0|_{\zeta=0} = -V_0|_{y=0}, \quad \lim_{\zeta\to\infty} P_0(\zeta,\theta,t) = 0$$

$$P_0|_{t=0} = 0,$$

where a and $\bar{\gamma}$ have the same meaning as in (2.8), but with $\phi$ replaced by $\theta$. The solutions of the problems (3.6), (3.9) are

$$(3.10)\qquad V_0(x,t) = \chi(x).\exp(-\gamma(x)t)$$

$$P_0(\zeta,\theta,t) = -\chi|_{\partial D}(\theta).\exp(-\bar{\gamma}(\theta)t).\frac{2}{\sqrt{\pi}}\int_{\zeta/\sqrt{4a(\theta)t}}^{\infty} e^{-\tau^2}\,d\tau$$

In the case $\chi = B_i^\varepsilon$ we proceed in a slightly different way, because $B_i^\varepsilon$ consists only of a boundary layer term along $\partial D$. We now take $V_0(.,0) \equiv 0$ and $P_0|_{y=0} \equiv 0$, $P_0|_{t=0} = b_i(\theta)\exp(-\mu(\theta)\zeta)$ with $\mu = \sqrt{\frac{\bar{\gamma}}{a}}$, see $(2.9)_b$. This leads us to

$$(3.10,\ \chi = B_i^\varepsilon)\qquad V_0 \equiv 0$$

$$P_0(\zeta,\theta,t) = b_i(\theta)e^{-\bar{\gamma}t}\int_0^\infty G(\zeta,\eta,t,\theta)e^{-\mu\eta}d\eta$$

with $G(\zeta,\eta,t,\theta) = \frac{1}{\sqrt{4\pi at}}\{\exp(-\frac{(\zeta-\eta)^2}{4at}) - \exp(-\frac{(\zeta+\eta)^2}{4at})\}$

It is a nice exercise in the use of the maximum principle for parabolic equations, [21], [22] to show that $|P_0| \leq |b_i| \max(e^{-\bar{\gamma}t}, e^{-\mu\zeta})$.

Note, that in (3.5) we can indeed take $\hat{\gamma} = \min \gamma(x)$.

Let us now exploit these results to find asymptotic approximations of the kernels $K_d^\varepsilon$ and $K_b^\varepsilon$ in $(3.4)_{d,b}$

$$(3.11)_d\quad K_d^\varepsilon = K_d + O(\sqrt{\varepsilon}e^{-\hat{\gamma}t}) \qquad\qquad (3.11)_b\quad K_b^\varepsilon = K_b + O(\sqrt{\varepsilon}e^{-\hat{\gamma}t})$$

with $[K_d](t) = e^{-\Gamma t}Z$, $Z_{k,i} = c_i(y_k)$, $\Gamma$ = the diagonal matrix with $\Gamma_{k,k} = \gamma(y_k)$ and $K_b \equiv 0$.

Using these approximations for the kernels the equations for $\xi^\varepsilon(t)$ reduce to

$$(3.12)_d\quad \xi(t) = e^{-\Gamma t}\eta_0 + \int_0^t e^{-\Gamma(t-\tau)}Z\xi(\tau)d\tau; \quad (3.12)_b\quad \xi = e^{-\Gamma t}\eta_0$$

We note, that $(3.12)_d$ is equivalent to a system of O.D.E.'s $\dot{\xi} = \Lambda\xi$, $\xi(0) = \eta_0$ with $\eta_0$ the vector with components $\psi(y_k)$ and $\Lambda = -\Gamma + Z$, as in $(2.12)_d$. Hence, the solution of $(3.12)_d$ is

$$(3.13)_d\quad \xi(t) = e^{\Lambda t}\eta_0$$

Substitution of these approximations in $(3.2)_{d,b}$ provides us with an approximation $\bar{u}$ of the solution $\bar{u}^\varepsilon$ of $(3.1)_{d,b}$. For the growth of $\bar{u}$ for $t \to \infty$ we find the following estimates

$(3.14)_d \quad |\bar{u}(x,t)| \leqslant C(\nu)e^{\nu t}$ $\qquad$ $(3.14)_b \quad |\bar{u}(x,t)| \leqslant C(\nu)e^{\nu t}$

$\nu > \max(-\hat{\gamma}, \mathrm{Re}\ \sigma(\Lambda)) = \nu_d^0$ $\qquad$ $\nu > -\hat{\gamma} = \nu_b^0$

with constants $C(\nu)$ independent of $\varepsilon$. Note, that this is in nice agreement with the results on the location of the spectrum of the controlled operator, compare (2.20).

In *the case of a 1st order unperturbed operator* as in $(1.4)_1$, the structure of an approximation of the solution of (3.3) is rather complicated. A sketch showing where the various layers are found, is given below.

fig. 2.

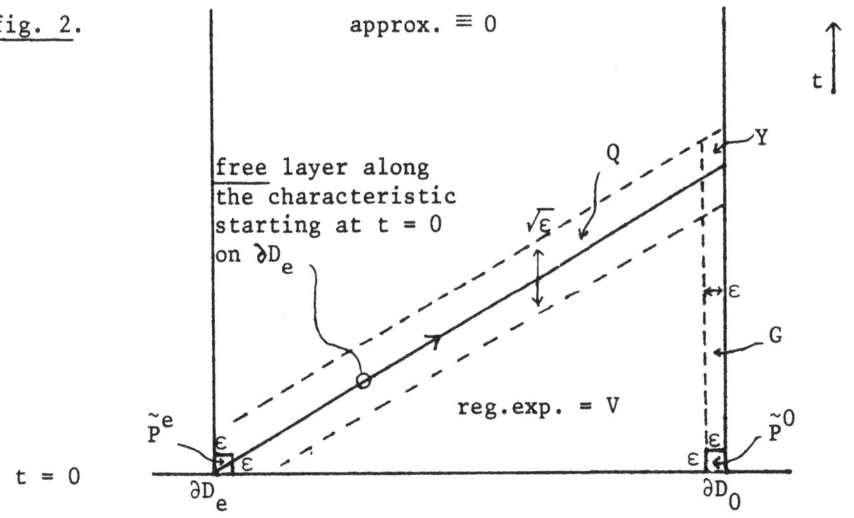


Our approximation now has the following form

$$(3.15) \qquad v^\varepsilon = \tilde{P}_0^e H^e + \tilde{P}_0^0 H^0 + (Q_0 + Y_0 H).H_-^c.H_+^c.(1 - H^e) +$$
$$+ (V_0 + G_0 H)(1 - H^e)(1 - H_-^c)(1 - H^0) + O(\varepsilon^{1/5} e^{-\omega t})$$

Here $H^e$, $H^0$, $H$ and $H_+^c$, $H_-^c$ denote suitable $C^\infty$ cut-off functions. Let $I(\alpha)$ be $\equiv 1$ for $\alpha \leqslant 1$ and $\equiv 0$ for $\alpha \geqslant 2$. Then we choose $H = I(T(\phi) - s)$, $H^e = I(s/\delta) \cdot I(t/\delta)$, $H^0 = I((T(\phi) - s)/\delta) . I(t/\delta)$, $H_+^c = I((t - s)/\delta^2)$

$H_-^c = I((s - t)/\delta^2)$ with $\delta = \varepsilon^{1/5}$. Of course, $T(\phi)$ and $s$ have the same meaning as in section 2. In the order of the error of the approximation $\omega \in \mathbb{R}$ can be taken arbitrary, as we shall see in the next section. The 0th order terms in (3.15) are found in the following way

(3.16) $$\frac{\partial V_0}{\partial t} = -\frac{\partial V_0}{\partial s} - \gamma V_0; \qquad V_0|_{t=0} = \chi$$

(3.17) $$\frac{\partial Q_0}{\partial s} = A\,\frac{\partial^2 Q_0}{\partial \tau^2} - \gamma Q_0$$

$$\lim_{\tau\to\infty} Q_0(s,\phi,\tau) = 0, \quad \lim_{\tau\to-\infty} Q_0(s,\phi,\tau) = V_0(s,\phi,s)$$

with $\tau = (t - s)/\sqrt{\varepsilon}$, $A = \sum_{i,j=1}^{n} \frac{\partial t_e}{\partial x_i} a_{ij} \frac{\partial t_e}{\partial x_i}$

(3.18) $$A^0\,\frac{\partial^2 Y_0}{\partial \zeta^2} - \frac{\partial Y_0}{\partial \zeta} = 0$$

$$Y_0(0,\phi,\tau) = -Q_0(T(\phi),\phi,\tau)$$

$$\lim_{\zeta\to\infty} Y_0(\zeta,\phi,\tau) = 0$$

with $\tau$ as above, $\zeta = (T(\phi) - s)/\varepsilon$ and $A^0 = A|_{\partial D_0}$. For the calculation of $\tilde{P}^e$ we introduce special coordinates near $\partial D_e$ in order to avoid singularities for $t \downarrow 0$ in $L_\varepsilon \tilde{P}^e$. Instead of $(s,\phi)$ we work with $(s,\theta)$ defined in such a way, that $\partial D_e = \{s = 0\}$ and:

$$L_2 = \bar{A}\,\frac{\partial^2}{\partial s^2} + \sum_{i,j=1}^{n-1} \bar{A}_{i,j}\,\frac{\partial^2}{\partial\theta_i\partial\theta_j} + \bar{A}_0\,\frac{\partial}{\partial s} + \sum_{i=1}^{n-1} \bar{A}_i\,\frac{\partial}{\partial\theta_i}$$

Such coordinates can be found by a procedure as in (3.7). Then, we define:

(3.19) $$\tilde{P}_0^e = \exp\left(-\tfrac{1}{2}\int_0^s \frac{\bar{A}_0(y,\theta)}{\bar{A}(y,\theta)}\,dy\right) . P_0^e(\eta,\theta,\tau_1)$$

with $\eta = s/\varepsilon$, $\tau_1 = t/\varepsilon$ and $P_0^e$ the solution of

(3.20) $$\frac{\partial P_0^e}{\partial \tau_1} = A^e\,\frac{\partial^2 P_0^e}{\partial \eta^2} - \frac{\partial P_0^e}{\partial \eta}$$

$$P_0^e(\eta,\theta,0) = \chi|_{\partial D_e}, \quad P_0^e(0,\theta,\tau) = 0$$

and $A^e = \bar{A}|_{\partial D_e} = A|_{\partial D_e}(\theta)$. In a completely analogous way we can introduce coordinates $s_0 = T(\phi) - s$ and $\bar{\theta}$ such that $\partial D_0 = \{s_0 = 0\}$ and

$$L_2 = \bar{A}' \frac{\partial^2}{\partial s_0^2} + \sum_{i,j=0}^{n-1} \bar{A}'_{i,j} \frac{\partial^2}{\partial \theta_i \partial \theta_j} + \bar{A}'_0 \frac{\partial}{\partial s_0} + \sum_{i=1}^{n-1} \bar{A}'_i \frac{\partial}{\partial \theta_i}$$

We put:

$$\tilde{P}_0^0 = \exp(-\tfrac{1}{2} \int_0^{s_0} \frac{\bar{A}'_0(y_0,\bar{\theta})}{\bar{A}'(y_0,\bar{\theta})} dy_0).\, P_0^0(\zeta,\bar{\theta},\tau_1) \tag{3.21}$$

with $\zeta = s_0/\varepsilon$ and $P_0^0$ the solution of

$$\frac{\partial P_0^0}{\partial \tau_1} = A^e \frac{\partial^2 P_0^0}{\partial \zeta^2} + \frac{\partial P_0^0}{\partial \zeta} \tag{3.22}$$

$$P_0^0(\zeta,\bar{\theta},0) = \chi|_{\partial D_0};\ P_0^0(0,\bar{\theta},\tau_1) = 0$$

with $A = A|_{\partial D_0}(\bar{\theta})$. In the case $\chi = B_i^\varepsilon$ the initial values for $P_0^0$ are taken as the 0th order term of the boundary layer expansion of $B_i^\varepsilon$, i.e. $\bar{G}_0(\zeta,\bar{\theta}) + \bar{B}_0|_{\partial D_0}(\bar{\theta})$ as given in $(2.17)_b$ with $b_0 = b_i|_{\partial D_0}$ and $b_e = b_i|_{\partial D_e}$. The solutions of these problems can easily be calculated.

$$V_0(s,\phi,t) = \chi(s - t,\phi)\exp(- \int_{s-t}^{s} \gamma(s',\phi)ds') \tag{3.23}$$

$$Q_0(s,\phi,\tau) = \exp(- \int_0^s \gamma(s',\phi)ds').q_0(r,\phi,\tau) \tag{3.24}$$

$$q_0(r,\phi,\tau) = \chi|_{\partial D_e}(\phi).\, \frac{1}{\sqrt{\pi}} \int_{\tau/\sqrt{4r}}^{\infty} e^{-t^2} dt$$

with $r = \int_0^s A(s',\phi)ds'$

$$Y_0(\zeta,\phi,\tau) = -Q_0(T(\phi),\phi,\tau).\exp(-\zeta/A^0) \tag{3.25}$$

$$P_0^e(\eta,\theta,\tau_1) = \chi|_{\partial D_e}(\theta).e^{(2\eta-\tau_1)/(4A^e)} \int_0^\infty G^e(\xi,\eta,\tau_1,\theta)e^{-\xi/(2A^e)} d\xi \tag{3.26}$$

with $G^e(\xi,\eta,\tau_1,\theta) = \frac{1}{\sqrt{4\pi A^e \tau_1}} \{\exp(- \frac{(\xi - \eta)^2}{4A^e\tau_1}) - \exp(- \frac{(\xi + \eta)^2}{4A^e\tau_1})\}$.

$$P_0^0(\zeta,\bar{\theta},\tau_1) = e^{-(2\zeta+\tau_1)/(4A^0)} \int_0^\infty G^0(\eta,\zeta,\bar{\theta},\tau_1)e^{-\eta/(2A^0)} P_0^0(\eta,\bar{\theta},0)d\eta \tag{3.27}$$

with $G^0(\eta,\zeta,\bar{\theta},\tau_1) = \frac{1}{\sqrt{4\pi A^0 \tau_1}} \{\exp(- \frac{(\xi - \eta)^2}{4A^0\tau_1}) - \exp(- \frac{(\zeta + \eta)^2}{4A^0\tau_1})\}$

Using the results as found in (3.15-27) we obtain the following approximations for the equations given in $(3.4)_{d,b}$

$$(2.28)_d \qquad \xi(t) = \eta(t) + \int_0^t K_d(\tau)\xi(t-\tau)d\tau;$$

$$(2.28)_b \qquad \xi(t) = \eta(t) + \frac{d}{dt} \{\int_0^t K_b(\tau)\xi(t-\tau)dt\}$$

with: $[K_d]_{k,i}(t) = c_i(s_k-t,\phi_k)\exp(-\int_{s_k-t}^{s_k}\gamma(s',\phi_k)ds')$ for $t < s_k$,

$= 0$ for $t > s_k$

$[K_b]_{k,i}(t) = 0$ for $t < s_k$,

$= b_{i,e}(\phi_k)\exp(-\int_0^{s_k}\gamma(s',\phi_k)ds')$ for $t > s_k$

and $\eta_k(t) = \psi(s_k-t,\phi_k)\exp(-\int_{s_k-t}^{s_k}\gamma(s',\phi_k))$ for $t < s_k$,

$= 0$ for $t > s_k$.

Here $(s_k,\phi_k)$ represents the observation points $y_k$ in $(s,\phi)$ coordinates. The errors in these approximations will be discussed in the next section. We observe, that for $t > s_k$ the k-th component of $(3.28)_d$ yields $\xi_k(t) = = \int_0^{s_k}[K_d(\tau)\xi(t-\tau)]_k d\tau$. It is also easy to check, that $(3.28)_b$ is equivalent to $\xi_k(t) = \eta_k(t) + [K_b(\infty)\xi(t-s_k)\check{H}(t-s_k)]_k$ with $\check{H}$ the Heaviside function. This illustrates that $(3.28)_{d,b}$ are equations with a finite *retardation*. Further, using the conditions which $\psi$ satisfies on $\partial D$, it is easy to check that $\eta$ and $\xi$ are continuous functions with bounded derivatives. With the exception of $t = s_k$ for $(3.28)_d$ and $t = ns_k$ for $(3.28)_b$ $\dot{\xi}$ and $\dot{\eta}$ are continuous. The equations $(3.28)_{d,b}$ can easily be solved using Laplace transformation. Let us denote the Laplace transform of f by Lf with $(Lf)(\lambda) = \int_0^\infty f(t)e^{-\lambda t}dt$. Because of the convolution structure of $(2.28)_{d,b}$ the equations for $L\xi$ are very simple:

$(3.29)_d$ $\quad (1 + \Omega_d(\lambda))L\xi = L\eta$ $\qquad\qquad (3.29)_b$ $\quad (1 - \Omega_b(\lambda)L\xi = L\eta$

with $\Omega_d$, $\Omega_b$ as in $(2.18)_{d,b}$. Hence:

$(3.30)_d$ $\quad \xi = L^{-1}[(1 + \Omega_d)^{-1}L\eta]$ $\qquad\qquad (3.30)_b$ $\quad \xi = L^{-1}[(1-\Omega_b)^{-1}L\eta]$

The inverse Laplace transformation is given by $(L^{-1}g)(t) = e^{\nu t}.(2\pi)^{-1}.$ $\int_{-\infty}^{\infty} g(\nu + i\eta)e^{i\eta t}d\eta$, see [23]. Here we can use any $\nu > \nu_d^1 =$ $= \max\{\mathrm{Re}\,\lambda |\, \omega_d(\lambda) = 0\}$, $\nu > \nu_b^1 = \max\{\mathrm{Re}\,\lambda |\, \omega_b(\lambda) = 0\}$ in $(3.30)_{d,b}$, respectively.

It is not difficult to verify, that substitution of $\xi$ in $(3.2)_{d,b}$ provides us with approximations $\bar{u}$ of the solutions $\bar{u}^\epsilon$ which satisfy the following estimates

(3.31)$_d$ $|\bar{u}(x,t)| \leqslant C(\nu)e^{\nu t}$, $\nu > \nu_d^1$ (3.31)$_b$ $|\bar{u}(x,t)| \leqslant C(\nu)e^{\nu t}$ $\nu > \nu_b^1$

This is in nice agreement with the asymptotic location of the spectrum as determined in section 2, compare (2.20).

## 4. On the asymptotic validity of the formal approximations

Here we shall derive some results on the correctness of the approximations of the solutions of the stationary problems, as found in section 2 and we shall also demonstrate, that the asymptotic location of the spectrum of the controlled operator given in that section is correct. Furthermore, we shall discuss the validity of our approximations of the solutions of the dynamic problems found in section 3.

Let us first consider *the case, where the unperturbed operator* $L_1$ *is of first order*. The following result will be very useful.

*Theorem:* Let $L_1$ be as in $(1.4)_1$ and let w be the solution of the dynamic problem

$$(4.1) \qquad \frac{\partial w}{\partial t} = L_\varepsilon w + R$$
$$w = 0 \text{ on } \partial D$$
$$w(.,0) = \psi \in C_0(\bar{D}) = \{\chi \in C(\bar{D}) \mid \chi = 0 \text{ on } \partial D\}$$

with $R \in C(\bar{D}\times(0,\infty))$ and bounded for $t \downarrow 0$.

Then:

$w \in C(\bar{D}\times[0,\infty))$ and $\exists \varepsilon_0, \beta, C, T > 0\ \forall \varepsilon \in (0,\varepsilon_0]\ \forall \omega \in [0,\frac{\beta}{\varepsilon}]\ \forall x \in \bar{D},\ t \geqslant 0$

$$(4.2) \qquad |w(x,t)| \leqslant Ce^{-\omega(t-T)}\max\{|\psi|_0, \sup_{[0,t]} |R(.,\tau)e^{\omega\tau}|_0\}$$

where $|\ |_0$ denotes the sup-norm for the domain D.

<u>*proof*</u>: We first take $\psi$, R smooth with $\mathrm{supp}(\psi) \subset D$ and $\mathrm{supp}(R) \subset \{t > 0\}$. Then, [24], there exists a solution w of (4.1) in $C^\infty(\bar{D}\times[0,\infty))$. The function $\tilde{w}$ defined by $w = \tilde{w}\exp(-\omega t)W(s)$ satisfies

$$(4.3) \qquad \frac{\partial \tilde{w}}{\partial t} = (\tilde{L}_\varepsilon + \omega)\tilde{w} + \tilde{R} \quad \text{with } \tilde{R} = Re^{\omega t}/W$$
$$\tilde{w} = 0 \text{ on } \partial D,\ w(.,0) = \tilde{\psi} = \psi/W$$

where the constant term of $\tilde{L}_\varepsilon + \omega$ equals $\tilde{\gamma} = W^{-1}(L_\varepsilon + \omega)W$. Now let W satisfy $(-\frac{\partial}{\partial s} + \alpha + \omega)W = -(1+\omega)W$, $W(0) = 1$ with $\alpha = \max \gamma$, i.e. $W(s) = \exp((\alpha + 2\omega + 1)s)$. It is easy to check, that $\tilde{\gamma} \leq -1-\omega+\tilde{C}\varepsilon(1+\alpha+2\omega)^2$; hence for $\omega \in [0,\beta/\varepsilon]$ with $\beta$ suitably chosen we have $\tilde{\gamma} \leq -\frac{1}{2}$. Using the maximumprinciple for parabolic equations, [22], [23], we obtain:

$$(4.4) \qquad \tilde{w}(x,t) \mp 2 \max\{|\tilde{\psi}|_0, \sup_{\tilde{t}\leq t} |\tilde{R}(.,\tilde{t})|_0\} \lesseqgtr 0$$

Of course, (4.4) implies (4.2).
The regularity assumptions on $\psi$, R can be weakened to those in the lemma by an approximation argument: $R_n \to R$ in $L^p(Q)$, $Q = D\times(0,T)$, p sufficiently large,using $L^p$ theory for parabolic equations, [24] and Sobolev's imbedding from $W^{1,p}(Q) \to C(\overline{Q})$, [25] and next $\psi_n \to \psi$ in $C_0(\overline{D})$. □

It is well-known, that $L_\varepsilon$ with Dirichlet boundary conditions generates an analytic semigroup $e^{L_\varepsilon t}$ on $C_0(\overline{D})$, [26]. A direct consequence of (4.2) is, that:

$$(4.5) \qquad |e^{L_\varepsilon t}|_0 \leq C \min(1, \exp(-\frac{\beta}{\varepsilon}(t - T)))$$

Note, that this estimate is in perfect agreement with the behaviour of the approximation for $e^{L_\varepsilon t}\chi$ given in (3.15). This is not completely trivial, since the free layer in (3.15) has a width $\sqrt{\varepsilon}$, but its erfc-structure makes it decay as in (4.5).
Using the characterisation of analytic semigroups in terms of the resolvent of the generator, [27], it is clear, that

$$(4.6) \qquad \sigma(L_\varepsilon) \subset \{\lambda | \operatorname{Re} \lambda \leq -\beta/\varepsilon\}$$

Since [18], [27], $(L_\varepsilon - \lambda)^{-1} = -\int_0^\infty e^{-\lambda t} e^{L_\varepsilon t} dt$ the resolvent satisfies the estimate:

$$(4.7) \qquad |(L_\varepsilon - \lambda)^{-1}|_0 \leq Ce^{\omega T}(\omega + \operatorname{Re} \lambda)^{-1}$$

for $\omega \in [0,\beta/\varepsilon]$, $\operatorname{Re} \lambda > -\omega$. This estimate in (4.5) is not only valid on $C_0(\overline{D})$, but on all of $C(\overline{D})$. This can be seen by using an $L^p$ approximation argument, $L^p$ theory for elliptic equations, [28] and Sobolev's imbedding from $W^{2,p}(D) \to C(\overline{D})$ for p sufficiently large. Let us now apply (4.5) to prove the validity of the approximations in $(2.14)_{d,b}$. We define $C' = \hat{C}_0 + (\hat{G}_0^0 + \varepsilon\hat{G}_1^0)H(T(\phi) - s)$, $B' = \overline{B}_0 + (\overline{G}_0^0 + \varepsilon\overline{G}_1^0)H(T(\phi) - s)$. Note, that compared with (2.14) we have included one more term in our boundary layer expansion. It is easy to verify that

$(4.8)_d$ $(L_\varepsilon - \lambda)(C^\varepsilon - C') = r_d$, $C^\varepsilon - C' = 0$ on $\partial D$ $\qquad$ $(4.8)_b$ $(L_\varepsilon - \lambda)(B^\varepsilon - B') = r_b$, $B^\varepsilon - B' = 0$ on $\partial D$

with $|r_d|_0 = O(\varepsilon)$, $|r_b|_0 = O(\varepsilon)$.
If $\lambda$ is fixed, then an immediate consequence of (4.7) is

$(4.9)_d$ $|C^\varepsilon - C'|_0 = O(\varepsilon)$ $\qquad$ $(4.9)_b$ $|B^\varepsilon - B'|_0 = O(\varepsilon)$

Therefore, the order of the error specified in (2.14) is indeed rigorous. Of course, also the order of the differences $\Omega_d^\varepsilon - \Omega_d$, $\Omega_b^\varepsilon - \Omega_b$ is then $O(\varepsilon)$.

Hence, if $\omega_d(\lambda) \neq 0$ $(,\omega_b(\lambda) \neq 0)$, we can invert $1 + \Omega_d^\varepsilon (, 1 - \Omega_b^\varepsilon)$ for $\varepsilon$ sufficiently small and the difference between $u^\varepsilon$ and its approximation u is rigorously $O(\varepsilon)$.
If $\omega_d(\lambda_k) = 0$ $(, \omega_b(\lambda_k) = 0)$ it follows from Rouche's theorem, [29], that $\det(1 + \Omega_d^\varepsilon(\lambda))$ $(, \det(1 - \Omega_b^\varepsilon(\lambda)))$ has $N \geqslant 1$ zero's in an $O(\varepsilon^{1/N})$ neighbourhood of $\lambda_k$ with N the order of the zero of $\omega_d(\lambda_k)(,\omega_b(\lambda_k))$. So, asymptotically close to $\lambda_k$ are points of the spectrum of the controlled operator. It is also easy to check, that if $\lambda(\varepsilon)$ is in the spectrum of the controlled operator and $\lim_{\varepsilon\downarrow 0} \lambda(\varepsilon) = \mu$ then $\mu$ is one of the zero's of $\omega_d(\lambda)(, \omega_b(\lambda))$. Hence, the zero's of $\omega_d(\lambda)(,\omega_b(\lambda))$ can be identified as the points of the spectrum of the controlled operator with a finite limit for $\varepsilon \downarrow 0$.

We can also use the above theorem to estimate rigorously the order of the error in our approximation (3.15) for the solution of the dynamic problem (3.3). We define: $v' = v + \varepsilon\{\tilde{P}_1^e H^e + \tilde{P}_1^0 H^0 + Y_1 H H_-^c H_+^c + G_1 H(1-H_-^c)(1-H^e)\}$ where $\tilde{P}_1^e$, $\tilde{P}_1^0$, $Y_1$, $G_1$ denote suitable next order terms in the various expansions. These corrections can be chosen in such a way that

$$(4.10)\quad \left(\frac{\partial}{\partial t} - L_\varepsilon\right)(v^\varepsilon - v') = \begin{cases} O(\varepsilon^{1/5}) & \text{for } 0 \leqslant t \leqslant T_1 \\ 0 & \text{for } t \geqslant T_1 \end{cases}$$

$$v^\varepsilon - v' = 0 \text{ on } \partial D$$

$$(v^\varepsilon - v')(.,0) = O(\varepsilon^{1/5})$$

Furthermore, our construction took care that the remainder terms in (4.10) satisfy the regularity conditions as required by (4.1), since we avoided unbounded singularities in $L_\varepsilon v'$ at $\partial D\times\{t=0\}$. The verification of the order functions in (4.10) is a long, paper-devouring business, but the calculations are rather straightforward. Now an application of (4.2) shows, that the order of the remainder in (3.15) is rigourous. Consequently it is clear, that the approximation of $K^\varepsilon_{d,b}$ by $K_{d,b}$, see (3.4), (3.28), takes place in the following sence:

$$(4.11)_d\quad |K^\varepsilon_d(t) - K_d(t)|_{k,i} \leqslant k_1(\omega)e^{-\omega t}\varepsilon^{1/5} + k_2\check{I}\left(\frac{t - s_k}{\varepsilon^{2/5}}\right)$$

$$(4.11)_b\quad |K^\varepsilon_b(t) - K_b(t)|_{k,i} \leqslant k_1\varepsilon^{1/5} + k_2\check{I}\left(\frac{t - s_k}{\varepsilon^{2/5}}\right)$$

Here $\check{I}$ denotes the indicator function of the interval $[-1,1]$. In the case of distributed control this is sufficient to show that $\xi$ approximates $\xi^\varepsilon$ in the following sense

$$(4.12)_d\quad |\xi^\varepsilon(t) - \xi(t)| \leqslant l(\nu)\varepsilon^{1/5}e^{\nu t}$$

where $\nu \in \mathbb{R}$ has to be $> \nu^1_d$, with $\nu^1_d$ as in (2.20). In order to show this we observe that $\xi^\varepsilon - \xi$ satisfies the equation

$$(4.13)_d\quad (I - K_d*)(\xi^\varepsilon-\xi) = \eta^\varepsilon-\eta+(K^\varepsilon_d-K_d)*\xi+(K^\varepsilon_d-K_d)*(\xi^\varepsilon-\xi)$$

where $*$ denotes the convolution operation. We consider this equation on the space $B_\nu = \{\xi \in \{C[0,\infty)\}^p \mid |\xi|_\nu = \sup|\xi(t)e^{-\nu t}| < \infty\}$ with $\nu > \nu^1_d$. We notice, that the equation $(I - K_d*)z = y$ is for $t \geqslant T = \max(s_k)$ equivalent to an autonomous retarded differential equation $\frac{d}{dt}(z-y) =$ $= L(z-y) + Ly$ with

$$(Lv)_k(t) = \sum_{i=1}^{p}\{[K_d]_{k,i}(s_k)v_i(t-t_k) - [K_d]_{k,i}(0)v_i(t) + \int_0^{s_k} (K_d)_{k,i}(\tau)v_i(t-\tau)d\tau\}$$

Using the theory of autonomous retarded differential equations, [30], it is clear that L generates a strongly continuous semigroup on $\{C[0,T]\}^p$, $T = \max s_k$ with $|e^{Lt}| \leq C(\nu')\exp(\nu' t)$ for each $\nu' > \nu_d^1$. The solution of $(I - K_d^*)z = y$ satisfies $z_t = e^{Lt}z_0 + y_t + \int_0^t e^{L(t-\tau)}Ly_\tau d\tau$ where $z_t \in \{C[0,T]\}^p$ is the element with $z_t(\bar\tau) = z(t+\bar\tau)$, $\bar\tau \in [0,T]$. It is now not difficult to show, that $I - K_d^*$ has a bounded inverse on $B_\nu$. Since $(K_d^\varepsilon - K_d)^*$ is an operator on $B_\nu$ with a norm of order $\varepsilon^{1/5}$ we can solve $(4.13)_d$ by a Neuman series and $(4.12)_d$ is a consequence from the fact $|\eta^\varepsilon - \eta + (K_d^\varepsilon - K_d)^*\xi|_\nu = O(\varepsilon^{1/5})$.

Using $(4.12)_d$ we find an approximation $\bar u$ of $\bar u_\varepsilon$ which satisfies

$$(4.14)_d \qquad |\bar u(.,t) - \bar u_\varepsilon(.,t)|_0 \leq \bar I(\nu)\varepsilon^{1/5}e^{\nu t} \qquad , \nu > \nu_d^1$$

This shows also that the spectrum of the controlled operator is in the half plane $\{\lambda | \operatorname{Re}\lambda \leq \nu\}$ if $\nu \in \mathbb{R}$, $\nu > \nu_d^1$ and $\varepsilon$ sufficiently small and further, that the analytic semigroup generated by $L_\varepsilon + \Pi_d$, [13], satisfies the estimate:

$$(4.15)_d \qquad |e^{(L_\varepsilon + \Pi_d)t}|_0 \leq C(\nu)e^{\nu t} \quad \text{for } \nu \in \mathbb{R}, \nu > \nu_d^1.$$

In the case of boundary control the estimate $(4.11)_b$ is not sufficient to carry out an analysis leading to something analogous to $(4.12)_d$, $(4.14)_b$, because of the derivative $\partial/\partial t$ in $(3.2)_b$ and $(3.4)_b$. Much more detailed information about the difference $K_b^\varepsilon - K_b$ and its time derivatives is necessary to do so. Our plan is to present such an analysis in a subsequent paper.

In *the case of a 0th order unperturbed operator* as in $(1.4)_0$ we can proceed analogously to above, see also [1]. Therefore we shall just state the results and leave the details of the derivations to the reader. For $L_0$ as in $(1.4)_0$ the solution w of (4.1) will be an element of $C(\bar D\times[0,\infty))$ satisfying an estimate

$$(4.16) \qquad |w| \leq C(\omega)e^{-\omega t}\max\{|\psi|_0, \sup_{[0,t]}|R(.,\tau)e^{\omega\tau}|_0\}$$

for each $\omega \in \mathbb{R}$, $-\omega \geq -\hat\gamma$. In combination with the selfadjointness of $L_\varepsilon$ this estimate yields: $\sigma(L_\varepsilon) \subset (-\infty,-\hat\gamma]$. The resolvent $(L_\varepsilon - \lambda)^{-1}$ satisfies the estimate:

$$(4.17)\qquad |(L_\varepsilon - \lambda)^{-1}|_0 \leq C(\mathrm{Re}\,\lambda + \hat{\gamma})^{-1} \qquad \text{for } \mathrm{Re}\,\lambda > -\hat{\gamma}$$

with a constant C independent of $\varepsilon$ and (4.17) is valid on all of $C(\bar{D})$. For Re $\lambda \leq -\hat{\gamma}$, Im $\lambda \neq 0$ we can also derive an estimate for $|(L_\varepsilon - \lambda)^{-1}f|_0$ with f sufficiently regular. To do this one starts with the observation $\|(L_\varepsilon - \lambda)^{-1}\|_{L_2} \leq |\mathrm{Im}\,\lambda|^{-1}$. Next using repeatedly a priori estimates for elliptic P.D.E!s, [28] and Sobolev's imbedding theorem from $W^{1,2}(D) \to C(\bar{D})$ for $1 > n/2$, [25] we end up with

$$(4.18)\qquad |(L_\varepsilon - \lambda)^{-1}f|_0 \leq C\,\frac{(1+|\lambda|)^k}{|\mathrm{Im}\,\lambda|}\,\varepsilon^{-k}|f|_{C^{2k}(\bar{D})}, \qquad 2k > \frac{n}{2}.$$

Including sufficiently many higher order terms in our expansions it is now easy to prove, that the orders of the remainders in $(2.6)_{d,b}$, $(2.11)_{d,b}$, (3.5) and $(3.11)_{d,b}$ are rigorously correct. Consequently if $\lambda \notin (-\infty,-\hat{\gamma}]$ and if in case of distributed control also $\lambda \notin \sigma(\Lambda)$, see $(2.12)_d$, the approximation u of $u^\varepsilon$ has an error $O(\sqrt{\varepsilon})$. In the case of distributed control the points $\sigma(\Lambda)\setminus(-\infty,-\hat{\gamma}]$ are exactly the finite limits in $\mathbb{C}\setminus(-\infty,-\hat{\gamma}]$ of eigenvalues of the controlled operator. In the case of boundary control the controlled operator has no finite limits of eigenvalues in $\mathbb{C}/(-\infty,-\hat{\gamma}]$. For the approximation of the dynamic solution in the case of distributed control we find

$$(4.19)_d\qquad |\bar{u}(.,t) - \bar{u}^\varepsilon(.,t)| \leq 1(\nu)\sqrt{\varepsilon}e^{\nu t}$$

for $\nu > \nu_d^0$. This shows that the spectrum of this controlled operator is for $\varepsilon$ sufficiently small in the half plane Re $\lambda \leq \nu$, $\nu > \nu_d^0$. The analytic semigroup generated by $L_\varepsilon + \Pi_d$, [13], satisfies the estimate

$$(4.20)\qquad |e^{(L_\varepsilon + \Pi_d)t}|_0 \leq C(\nu)e^{\nu t} \qquad \text{for } \nu \in \mathbb{R},\ \nu > \nu_d^0.$$

Let us conclude this section with a few remarks about stabilizability of the system. Considering the results on the asymptotic location of the spectrum of the controlled operator we can not hope that feed back as in $(1.1)_{d,b}$ will improve the stability properties of the uncontrolled system, if $\varepsilon$ is sufficiently small. From the point of view of general stabilizability results this is at a first glance somewhat puzzling. Using [14], [2] we know, that in the case of $(1.4)_0$ with a 1 dimensional domain, when all eigenvaules of $L_\varepsilon$ are simple, it is possible, for a fixed $\varepsilon$, to choose 1 observation point $y_1^\varepsilon$ and 1 distributed control input $c_1^\varepsilon$, such that $\sigma(L_\varepsilon + \Pi_d) \subset (-\infty,-\alpha]$, where $\alpha$ can be prescribed arbitrary. But, the reason why a control (1.1) is not suitable to do this is rather transparent: we assumed that the input functions and observation points do not depend wildly on $\varepsilon$ for $\varepsilon \downarrow 0$. Hence, for a fixed $\alpha > \hat{\gamma}, c_1^\varepsilon$, $y_1^\varepsilon$ must have a wildly fluctuating structure for $\varepsilon \downarrow 0$.

This can also be seen from the construction of the stabilizing control in [14], [2]. In the next section we shall see, that controls as in (1.1) can be used to optimize a different kind of performance index of the system.

## 5. An example of near optimal feed-back control

Let us consider the following controlled problem

$$\text{(5.1)} \qquad \frac{\partial u}{\partial t} = L_\varepsilon u + \tilde{\Pi}_d u + h$$

$$u = 0 \text{ on } \partial D,\ u(.,0) = \psi$$

with $\tilde{\Pi}_d u = c_0 + c(\delta_y u - i)$ and $h = f_0 + \underline{\omega} f$. Note that the control consists of a permanent part $c_0$ and a feed back part based on the comparison of the observation of u in the point y with the reference value i. We suppose that h is a stationary, autonomous inhomogenity, which, in various situations where the system is likely to operate, has a distribution $f_0 + \underline{\omega} f$, $f_0$, $f \in C^\infty(\bar{D})$ and $\underline{\omega}$ a stochastic parameter with $E(\underline{\omega}) = 0$. Now we want to determine $c_0$, c, i and y in such a way, that the expected costs J are minimal, with J the following quadratic functional

$$\text{(5.2)} \qquad J = E\{ \int_D [(u_{stat} - g)^2 + \mu_0 c_0^2 + \mu c^2(\delta_y u_{stat} - i)^2] dx\}$$

under the obvious side condition that the stationary solution $u_{stat}$ of (5.1) is exponentially stable:

(5.3) $\qquad |u(.,t) - u_{stat}|_0 \leqslant C(\nu).|u_{stat} - \psi|_0.e^{\nu t}$ with $-\hat{\gamma} < \nu < 0.$

In (5.2), $g \in C^\infty(\bar{D})$ has the interpretation of the ideal stationary state of the system, $\mu_0$ and $\mu$ are constants $> 0$. Using the results of the previous sections we are able to solve this optimization problem in an approximate sense for $\varepsilon \downarrow 0$. For simplicity we consider the case $(1.4)_0$ with : $\gamma \equiv \hat{\gamma}$.

In order to do so we first note, that

(5.4) $\qquad u_{stat} \simeq \gamma^{-1}\{\alpha c + c_0 + f_0 + \underline{\omega}(\beta c + f)\} + \ldots$

with $\alpha = [\gamma i - c_0(y) - f_0(y)]/[c(y) - \gamma]$, $\beta = -f(y)/[c(y) - \gamma]$.

The remainder term in (5.4) only contributes as $O(\sqrt{\varepsilon})$ to the expected costs. Hence, J is approximately given by

(5.5) $\qquad J = J_0 + \int_D \{A_0 c_0^2 + 2B_0 c_0 + Ac^2 + 2Bc + 2Ecc_0\}dx + O(\sqrt{\varepsilon})$

with $J_0$ the expected costs in the uncontrolled situation, $A_0 = \gamma^{-2} + \mu_0$, $B_0 = \gamma^{-1}f_0 - g$, $A = (\gamma^{-2} + \mu)(\alpha^2 + \omega_2\beta^2)$, $B = \alpha\gamma^{-1}B_0 + \omega_2\gamma^{-2}\beta f$, $E = \alpha\gamma^{-2}$, where $\omega_2 = E(\underline{\omega}^2)$, the second moment of $\underline{\omega}$.

Let us first minimize the second term $J_1$ in (5.5) without consideration of the stability condition (5.3). Note, that

$J_1 = \int_D\{A_0(c_0 + B_0A_0^{-1} + EA_0^{-1}c)^2 + \hat{A}_0(c + \hat{B}_0\hat{A}_0^{-1})^2\}dx - \int_D \{A_0^{-1}B_0^2 + \hat{A}_0^{-1}\hat{B}_0^2\}dx$

$J_1$ is minimal, if

(5.6) $\qquad c_0 = -A_0^{-1}(B_0 + Ec)$

$\qquad\qquad c = -\hat{A}_0^{-1}\hat{B}_0$

with $\hat{A}_0 = \alpha^2 A_1 + \beta^2 A_2$, $\quad A_1 = \gamma^{-2}\{(1 + \mu\gamma^2) - (1 + \mu_0\gamma^2)^{-1}\} > 0$

$\qquad\qquad A_2 = \gamma^{-2}(1 + \mu\gamma^2)\omega_2 > 0$

$\hat{B}_0 = \alpha B_1 + \beta B_2$, $\quad B_1 = \gamma^{-1}B_0\{1 - \gamma(1 + \mu_0\gamma^2)^{-1}\}$

$\qquad\qquad B_2 = \gamma^{-2}\omega_2 f$

and the value of the minimum is

(5.7) $\qquad J_1^{min} = J_1^0 - (\alpha^2 A_1 + \beta^2 A_2)^{-1}(\alpha^2 N_1 + 2\alpha\beta M + \beta^2 N_2)$

with $J_1^0 = -A_0^{-1}\|B_0\|^2$, $N_1 = \|B_1\|^2$, $N_2 = \|B_2\|^2$, $M = \langle B_1, B_2 \rangle$ where $\| \;\|$ and $\langle\, ,\, \rangle$ denote the norm and innerproduct on $L_2(D)$. $J_1^0$ is independent of the choice of $\alpha, \beta$, but the second term in (5.7) is still a function of $z = \beta/\alpha$. If $M \neq 0$ the best choice for z is

$$(5.8) \qquad z = z_0 = (2m)^{-1}\{(a - n) + \mathrm{sign}(m)\sqrt{(a - n)^2 + 4am^2}\}$$

with $a = A_1/A_2$, $n = N_1/N_2$ and $m = M/N_2$.

Let us now think about the stability condition (5.3). Because of $(4.20)_d$ in combination with (2.20) this condition is satisfied if

$$(5.9) \qquad c(y) - \gamma < \nu$$

Now our *near optimal control* is constructed in the following way (i) we choose y in the interior of D such that $f(y) \neq 0$ (ii) we construct $\tilde{c}$ = the near optimal c in such a way, that $\tilde{c}(y) = -1$ (iii) as a consequence we must have $\beta = f(y)/(\gamma + 1)$ and we choose $\alpha = \beta/z_0$ (iv) we take $\tilde{c}$ almost as in (5.6), namely

$$(5.10) \qquad \tilde{c} = -\alpha^{-1}(A_1 + z_0^2A_2)^{-1}(B_1 + z_0B_2) + [\alpha^{-1}(A_1 + z_0^2A_2)^{-1}(B_1 + z_0B_2)(y) - 1]\rho\left(\frac{x - y}{\delta}\right)$$

with $\rho$ a smooth function with compact support and $\rho \equiv 1$ on a neighbourhood of $0 \in \mathbb{R}$ and $\delta$ an $o(1)$ function, which we shall presently relate to $\varepsilon$ (v) we then obtain $c_0$ as $-A_0^{-1}(B_0 + E\tilde{c})$, see (5.6) (vi) finally, our choice of i is determined by the consistency condition $\beta/\alpha = z_0 = -f(y)/(\gamma i - c_0(y) - f_0(y))$, i.e.

$$(5.11) \qquad i = (\gamma z_0)^{-1}\{z_0(c_0(y) + f_0(y)) - f(y)\}$$

Note, that our asymptotic analysis given in the previous sections remains valid for control inputs with a structure as in (5.10), though the order of the remainders increase to $O(\varepsilon\delta^{-2} + \sqrt{\varepsilon})$. It is easy to check, that this nearoptimal control produces costs which are at most $(\varepsilon^{\frac{1}{2}} + \varepsilon\delta^{-2} + \delta^n)$ above the genuin minimum. Hence, $\delta = \varepsilon^{1/3}$ for $n = 1$, $= \varepsilon^{\frac{1}{4}}$ for $n \geqslant 2$ is a good choice for $\delta$. The analysis of the non-generic case, where $M = 0$, is left to the reader as an exercise.

References

[ 1] van Harten, A., 1979
"Feedback control of singularly perturbed heating problems",
Lecture Notes in Math., 711, Springer, Berlin.

[ 2] Curtain, R.F., Pritchard, A.J., 1978
"Infinite dimensional linear systems theory",
Lecture Notes in Contr. and Inf. Sc., 8, Springer, Berlin.

[ 3] Owens, D.H., 1980
"Spatial kinetics in nuclear reactor systems", in:
"Modelling of dynamical systems", P. Peregrinus, Stevenage.

[ 4] Schumacher, J.M., 1981
"Dynamic feedback in finite- and infinite dimensional linear systems",
Math. Centre, A'dam.

[ 5] Triggiani, R., 1979
"On Nambu's boundary stabilizability problem for diffusion processes",
J. Diff. Eq., 33.

[ 6] Eckhaus, W., 1979
"Asymptotic analysis of singular perturbations",
North-Holland, A'dam.

[ 7] Fife, P.C., 1974
"Semi-linear elliptic boundary value problems with a small parameter",
Arch. Rat. Mech., 52.

[ 8] van Harten, A., 1975
"Singularly perturbed non-linear 2nd order elliptic boundary value problems",
thesis, Math. Inst., Utrecht.

[ 9] van Harten, A., 1978
"Non-linear singular perturbation problems: proofs of correctness",
J. Math. An. Appl., 65.

[10] de Groen, P.P.N., 1976
"Singularly perturbed differential operators of second order",
Math. Centre, A'dam.

[11] Besjes, J.G., 1974
"Singular perturbation problems for linear parabolic differential operators of arbitrary order",
J. Math. An. Appl. 48

[12] van Harten, A., Schumacher, J.M., 1980
Well-posedness of some evolution problems in the theory of automatic feedback control for systems with distributed parameters",
SIAM. J. Contr.Opt., 18.

[13] van Harten, A., 1979
"On the spectral properties of a class of elliptic FDE arising in feedback control theory for diffusion processes",
preprint nr. 130, Math. Inst., Un. of Utrecht.

[14] Triggiani, R., 1975
"On the stabilization problem in Banach spaces",
J. Math. An. Appl., 52.

[15] Balas, M.J., 1979
"Feedback control of linear diffusion processes",
Int. J. Control, 29.

[16] Balas, M.J., 1982
"Reduced order feedback control of distributed parameter systems via singular perturbation methods",
J. Math. An. Appl., 87.

[17] Triggiani, R., 1980
"Boundary feedback stabilizability of parabolic equations",
Appl. Math. and Opt., 6.

[18] Kato, T., 1966
"Perturbation theory for linear operators",
Springer, Berlin.

[19] Wilkinson, J.H., 1965
"The algebraic eigenvalue problem",
Oxford Un. Press.

[20] Lions, J.L., 1973
"Perturbations singulières dans les problèmes aux limites et en contrôle optimal",
Lect. Notes in Math., 323, Springer, Berlin.

[21] Friedman, A., 1976
"Stochastic differential equations and applications",
Ac. Press, New York.

[22] Protter, M.H., Weinberger, H.F., 1967
"Maximum principles in differential equations",
Prentice Hall, London.

[23] Schwartz, L., 1965
"Méthodes mathématiques poûr les sciences physiques",
Hermann, Paris.

[24] Ladyzenskaja, D.A., Solonnikov, V.A., Uraltseva, N.N., 1968
"Quasi-linear equations of parabolic type",
Am. Math. Soc. Transl., 23.

[25] Adams, R.A., 1975
"Sobolev spaces",
Ac. Press, New York.

[26] Stewart, H.B., 1980
"Generation of semigroups by strongly elliptic operators under general boundary conditions",
Trans. A.M.S., 259.

[27] Krasnoselskii, M.A., etal., 1976
"Integral operators in spaces of summable functions",
Noordhoff, Leyden.

[28] Agmon, S., Douglis, A., Nirenberg, N., 1959
"Estimates near the boundary for solutions of elliptic PDE satisfying general BC",
Comm. Pure Appl. Math., 12.

[29] Conway, J.B., 1973
"Functions of one complex variable",
Springer, Berlin.

[30] Hale, J., 1971
"Functional differential equations",
Springer, Berlin.

Printed in the United States
By Bookmasters